AF361871

Advanced Nanocellulose-Based Materials: Production, Properties and Applications

Advanced Nanocellulose-Based Materials: Production, Properties and Applications

Editors

Carla Vilela
Carmen S. R. Freire

MDPI • Basel • Beijing • Wuhan • Barcelona • Belgrade • Manchester • Tokyo • Cluj • Tianjin

Editors
Carla Vilela
Department of Chemistry,
CICECO - Aveiro Institute of
Materials
University of Aveiro
Aveiro
Portugal

Carmen S. R. Freire
Department of Chemistry,
CICECO - Aveiro Institute of
Materials
University of Aveiro
Aveiro
Portugal

Editorial Office
MDPI
St. Alban-Anlage 66
4052 Basel, Switzerland

This is a reprint of articles from the Special Issue published online in the open access journal *Nanomaterials* (ISSN 2079-4991) (available at: www.mdpi.com/journal/nanomaterials/special_issues/nanocellulose_production_applications).

For citation purposes, cite each article independently as indicated on the article page online and as indicated below:

LastName, A.A.; LastName, B.B.; LastName, C.C. Article Title. *Journal Name* **Year**, *Volume Number*, Page Range.

ISBN 978-3-0365-3191-5 (Hbk)
ISBN 978-3-0365-3190-8 (PDF)

Contents

About the Editors

Carla Vilela

Dr. Carla Vilela (ORCID: 0000-0002-9212-2704) is a Principal Researcher at CICECO Aveiro Institute of Materials (University of Aveiro, Portugal), where she coordinates the research line L3—Sustainability. Her main research interest includes the sustainable use of biopolymers, namely polysaccharides (e.g., nanocellulose, pullulan, chitosan), for the development of novel functional nanostructured materials for both technological (e.g., active food packaging and fuel cells) and biomedical (e.g., drug delivery and wound healing) applications. She is the author of 71 SCI papers (h-index: 25; citations: 2406), 2 books, 5 book chapters, 2 filed patents, 10 conference proceedings, and more than 100 communications in national and international conferences.

Carmen S. R. Freire

Dr. Carmen Freire (ORCID: 0000-0002-6320-4663) is a Principal Researcher with Habilitation at CICECO—Aveiro Institute of Materials (University of Aveiro, Portugal) in the biorefinery and bio-based materials research area. Her research interests include the production and application of biogenic nanofibers (bacterial cellulose and protein fibrils), nanostructured biocomposites and hybrid materials; bio-based materials for biomedical applications (wound healing, drug delivery and 3D bioprinting), biocomposites and functional paper materials; the chemical modification of (nano)cellulose fibers and other polysaccharides and their characterization and applications. CFreire is the author of more than 240 papers (with +9800 citations, h-index 56), 7 patents, 9 book chapters and more than 200 communications in international and national conferences.

 *nanomaterials*

MDPI

Editorial

Advanced Nanocellulose-Based Materials: Production, Properties, and Applications

Carmen S. R. Freire  and Carla Vilela *

CICECO—Aveiro Institute of Materials, Department of Chemistry, University of Aveiro, 3810-193 Aveiro, Portugal; cfreire@ua.pt
* Correspondence: cvilela@ua.pt

Citation: Freire, C.S.R.; Vilela, C. Advanced Nanocellulose-Based Materials: Production, Properties, and Applications. *Nanomaterials* **2022**, *12*, 431. https://doi.org/10.3390/nano12030431

Received: 12 January 2022
Accepted: 18 January 2022
Published: 27 January 2022

Publisher's Note: MDPI stays neutral with regard to jurisdictional claims in published maps and institutional affiliations.

Natural polymers, such as polysaccharides and proteins, are being extensively utilized as substrates to create advanced materials [1–4]. Within the vast portfolio of natural polymers, bacterial nanocellulose (BNC), cellulose nanofibers (CNFs), and cellulose nanocrystals (CNCs), viz. the three nanometric forms of cellulose, are currently at the spotlight in numerous fields of modern science and technology [5–7]. The eco-friendly nature, peculiar features and multiple functionalities of these nanoscale cellulosic substrates are being investigated to engineer advanced nanocomposites and nanohybrid materials for application in manifold domains, such as mechanics, optics, electronics, energy, environment, biology, and medicine.

This Special Issue of *Nanomaterials* entitled Advanced Nanocellulose-Based Materials: Production, Properties, and Applications, brings together a compilation of original research and review contributions from world-leading scientists working with nanocellulose. Hence, this Special Issue contains a collection of one review paper about the characterization of cellulose nanomaterials [8] and eight research papers focused on the use of BNC [9–11], CNFs [12–15], and CNCs [16] as reinforcements in composites [13–15] and to produce ion-exchange membranes for fuel cells [9], patches for tissue engineering and wound healing [10,11], and nanosystems or nanocarriers for cancer treatment [15,16].

Under the title *Recent Progress on the Characterization of Cellulose Nanomaterials by Nanoscale Infrared Spectroscopy*, Zhu et al. [8] reviewed the latest advances in the applications of current state-of-the-art nanoscale infrared spectroscopy and imaging techniques, namely atomic force microscope-based infrared spectroscopy (AFM-IR) and infrared scattering scanning near-field optical microscopy (IR s-SNOM), to characterize cellulose nanomaterials. As stated by the authors, AFM-IR and IR s-SNOM are two techniques for compositional analysis and chemical mapping at the nanoscale spatial resolution that can also deliver insightful information on mechanical, thermal, and electrical properties for cellulose nanomaterials [8].

The study by Vilela et al. [9] demonstrated the feasibility of combining BNC (i.e., a microbial exopolysaccharide) with a water-soluble anionic sulfonated lignin derivative (i.e., lignosulfonates) and a natural crosslinker (i.e., tannic acid) to produce freestanding homogeneous membranes with good mechanical performance (maximum Young's modulus of ca. 8.2 GPa) and moisture-uptake capacity (ca. 78% after 48 h) and a maximum ionic conductivity of 23 mS cm^{-1} (at 94 °C and 98% relative humidity). Even though the conductivity values achieved are comparable or higher than other fully biobased ion-exchange membranes reported in the literature, they are still two orders of magnitude lower than the standard commercial Nafion^{TM} ionomer currently in use in fuel cells. Still, and according to the authors, this study might contribute to the long and laborious path towards the development of eco-friendly conducting separators, especially through the exploitation of surplus raw materials from agricultural and industrial by-products [9].

Similarly interesting is the investigation of Kutová et al. [10] that examined the effect of the drying method (air-drying or freeze-drying) and subsequent argon plasma modification of BNC on the adhesion and growth of human keratinocyte cells (HaCaT cell line). The air-dried method followed by argon plasma modification yielded a BNC membrane with lower porosity (total pore volume (V_p) = 0.142 $\pm$ 0.008 cm^3 g^{-1}) and specific surface area (S_{BET} = 140.5 $\pm$ 4.8 m$^2\cdot$g^{-1}) when compared with the freeze-dried and argon plasma-modified BNC (V_p = 0.308 $\pm$ 0.015 cm^3 g^{-1} and S_{BET} = 156.3 $\pm$ 1.5 m$^2\cdot$g^{-1}). The growth and adhesion of HaCaT cells, when BNC was air-dried or freeze-dried, was markedly improved by the argon plasma modification. As stated by the authors, the surface modification rendered BNC a good support for the adhesion, growth, and viability of human HaCaT keratinocytes and thus can be considered highly promising materials for skin tissue engineering and wound healing [10].

In a different study, Fonseca et al. [11] reported the preparation and characterization of multilayered patches composed of oxidized BNC membranes (prepared via TEMPO (2,2,6,6-tetramethylpiperidin-1-yl)oxyl)-mediated oxidation of BNC) loaded with dexpanthenol (DEX), an active ingredient used in topical products for the treatment of dermatological conditions (e.g., wound healing). Spin-assisted layer-by-layer (LbL) assembly was employed to coat the DEX-loaded BNC membranes with alternate layers of chitosan and alginate polyelectrolytes, yielding multilayered patches with up to a total of 21 layers. The patches retarded the release of the DEX from the three-dimensional nanostructure of BNC, depending on the number of layers (ca. 95% after 16 h for the DEX-loaded BNC membrane versus ca. 65% after 90 h for the DEX-loaded BNC membrane with 21 layers). Furthermore, the multilayered patches inhibited the growth of *Staphylococcus aureus* (owing to the antimicrobial action of chitosan), were non-cytotoxic to human keratinocyte cells (HaCaT cell line) and showed good migration results in the wound healing scratch assay, suggesting the pertinence of these systems in the treatment of skin wounds [11].

In the domain of CNFs, and as highlighted in the research by Jonasson et al. [12], the lignin content influenced the one-pot direct TEMPO-oxidative) nanofibrillation of wood cell wall. In fact, wood from poplar trees with high lignin content (30.0 wt.%) was more easily fibrillated as revealed by the higher nanofibril yield (68% and 45%) and suspension viscosity (27 and 15 mPa·s) than low lignin wood (17.4 and 19.7 wt.%). Moreover, the surface area (114 m$^2\cdot$g^{-1}) and pore size (5.0 nm) of the oxidized high lignin wood were higher than those of low lignin wood (76 m$^2\cdot$g^{-1} and 4.4 nm, respectively), which means that porosity is a factor that can also beneficially impact the isolation of CNFs from wood [12].

Another original work was reported by Nissilä et al. [13], who investigated the use of ice-templated CNF filaments as a reinforcement material in epoxy resin-based composites. The authors successfully manufactured cellulose nanocomposites with an oriented structure and a strong fiber–matrix interface by preparing unidirectionally aligned CNF-filament mats via ice-templating followed by chemical vapor deposition to obtain silane-treated CNF filaments. The impregnation of these mats with a bio-epoxy resin through vacuum infusion originated composites with 18 wt.% fiber content oriented along the freezing direction of the ice crystals, which translated into composite materials with improved mechanical performance (storage modulus) up to 2.5-fold [13].

Xue et al. [14] prepared high-strength regenerated cellulose composite fibers reinforced with CNFs and nanosilica (nano-SiO$_2$). The incorporation of 1% of CNFs and 1% of nano-SiO$_2$ into a cellulose solution of an ionic liquid, 1-allyl-3-methylimidazolium chloride (AMIMCl), improved the mechanical properties of the regenerated composite cellulose fibers by 47.46%. In addition, the viscosity of the cellulose/AMIMCl mixtures (with or without CNFs and nano-SiO$_2$) was characteristic of pseudoplastic fluids, and the storage and loss moduli (elasticity and viscous moduli) decreased with the temperature, representing a reduction in viscoelasticity [14].

Equally interesting is the study of Yusefi et al. [15], which investigated the use of the LbL methodology to assemble nanocomposites (diameter below 50 nm) of chitosan reinforced with CNFs for the encapsulation of an anticancer drug, i.e., 5-fluorouracil. The nanocomposites, with a pH-responsive behavior and non-cytotoxicity towards the human colonic CCD112 cell line, exhibited a drug encapsulation efficiency of ca. 86% and were able to eliminate (as intended) 56.42 $\pm$ 0.41% of the human colorectal cancer HCT116 cells at a concentration of 250 µg mL^{-1}, showing their potential as nanocarriers of anticancer drugs [15].

Lastly, Pinto et al. [16] developed CNC nanosystems functionalized with a chitosan derivative holding both targeting (folic acid) and imaging (fluorescein isothiocyanate) functions. The simple and environmentally friendly method of physical adsorption was adopted, considering the opposite charges of CNCs with a zeta (ζ)-potential of -11.5 ± 0.7 mV at pH 7 and the chitosan derivative with a ζ-potential value of $+51.9 \pm 3.6$ mV at pH 3. The CNCs nanosystems demonstrated improved internalization (up to five-fold) in human breast adenocarcinoma cells (MDA-MB-231) and a potential anti-proliferative effect, hinted by exometabolomic analysis, suggesting the applicability of these nanosystems in active targeted cancer therapy [16].

Overall, the nine papers in this Special Issue of *Nanomaterials*—belonging to the section Nanocomposite Materials—addressed the exploitation of nanocellulosic substrates to manufacture advanced nanocomposites and nanohybrid materials for both biomedical (e.g., wound healing [11] and anticancer treatment [15,16]) and technological (e.g., fuel cells [9]) applications.

Funding: This work was developed within the scope of the project CICECO—Aveiro Institute of Materials (UIDB/50011/2020 & UIDP/50011/2020), financed by national funds through the Portuguese Foundation for Science and Technology (FCT)/MCTES. FCT is also acknowledged for the re-search contract under Scientific Employment Stimulus to C.S.R.F. (CEECIND/00464/2017) and C.V. (CEECIND/00263/2018 and 2021.01571.CEECIND).

Acknowledgments: The Guest Editors wish to acknowledge the authors for their contributions to this Special Issue, the reviewers for their challenging work in reviewing the submitted papers, and the editorial staff of *Nanomaterials* for their exceptional support.

Conflicts of Interest: The authors declare no conflict of interest.

References

1. Silva, A.C.Q.; Silvestre, A.J.D.; Vilela, C.; Freire, C.S.R. Natural Polymers-based Materials: A Contribution to a Greener Future. *Molecules* **2022**, *27*, 94. [CrossRef] [PubMed]
2. Silva, N.H.C.S.; Vilela, C.; Marrucho, I.M.; Freire, C.S.R.; Pascoal Neto, C.; Silvestre, A.J.D. Protein-based materials: From sources to innovative sustainable materials for biomedical applications. *J. Mater. Chem. B* **2014**, *2*, 3715–3740. [CrossRef] [PubMed]
3. Vilela, C.; Figueiredo, A.R.P.; Silvestre, A.J.D.; Freire, C.S.R. Multilayered materials based on biopolymers as drug delivery systems. *Expert Opin. Drug Deliv.* **2017**, *14*, 189–200. [CrossRef] [PubMed]
4. Vilela, C.; Pinto, R.J.B.; Pinto, S.; Marques, P.A.A.P.; Silvestre, A.J.D.; Freire, C.S.R. *Polysaccharide Based Hybrid Materials: Metals and Metal Oxides, Graphene and Carbon Nanotubes*, 1st ed.; Springer Nature: Basel, Switzerland, 2018; ISBN 978-3-030-00346-3.
5. Carvalho, J.P.F.; Silva, A.C.Q.; Silvestre, A.J.D.; Freire, C.S.R.; Vilela, C. Spherical Cellulose Micro and Nanoparticles: A Review of Recent Developments and Applications. *Nanomaterials* **2021**, *11*, 2744. [CrossRef] [PubMed]
6. Vilela, C.; Silvestre, A.J.D.; Figueiredo, F.M.L.; Freire, C.S.R. Nanocellulose-based materials as components of polymer electrolyte fuel cells. *J. Mater. Chem. A* **2019**, *7*, 20045–20074. [CrossRef]
7. Almeida, T.; Silvestre, A.J.D.; Vilela, C.; Freire, C.S.R. Bacterial nanocellulose toward green cosmetics: Recent progresses and challenges. *Int. J. Mol. Sci.* **2021**, *22*, 2836. [CrossRef] [PubMed]
8. Zhu, Q.; Zhou, R.; Liu, J.; Sun, J.; Wang, Q. Recent progress on the characterization of cellulose nanomaterials by nanoscale infrared spectroscopy. *Nanomaterials* **2021**, *11*, 1353. [CrossRef] [PubMed]
9. Vilela, C.; Morais, J.D.; Silva, A.C.Q.; Muñoz-Gil, D.; Figueiredo, F.M.L.; Silvestre, A.J.D.; Freire, C.S.R. Flexible nanocellulose/lignosulfonates ion conducting separators for polymer electrolyte fuel cells. *Nanomaterials* **2020**, *10*, 1713. [CrossRef] [PubMed]
10. Kutová, A.; Staňková, L.; Vejvodová, K.; Kvítek, O.; Vokatá, B.; Fajstavr, D.; Kolská, Z.; Brož, A.; Bačáková, L.; Švorčík, V. Influence of drying method and argon plasma modification of bacterial nanocellulose on keratinocyte adhesion and growth. *Nanomaterials* **2021**, *11*, 1916. [CrossRef] [PubMed]
11. Fonseca, D.F.S.; Carvalho, J.P.F.; Bastos, V.; Oliveira, H.; Moreirinha, C.; Almeida, A.; Silvestre, A.J.D.; Vilela, C.; Freire, C.S.R. Antibacterial multi-layered nanocellulose-based patches loaded with dexpanthenol for wound healing applications. *Nanomaterials* **2020**, *10*, 2469. [CrossRef] [PubMed]
12. Jonasson, S.; Bünder, A.; Berglund, L.; Hertzberg, M.; Niittylä, T.; Oksman, K. The Effect of High Lignin Content on Oxidative Nanofibrillation of Wood Cell Wall. *Nanomaterials* **2021**, *11*, 1179. [CrossRef] [PubMed]

13. Nissilä, T.; Wei, J.; Geng, S.; Teleman, A.; Oksman, K. Ice-templated cellulose nanofiber filaments as a reinforcement material in epoxy composites. *Nanomaterials* **2021**, *11*, 490. [CrossRef] [PubMed]
14. Xue, Y.; Qi, L.; Lin, Z.; Yang, G.; He, M.; Chen, J. High-strength regenerated cellulose fiber reinforced with cellulose nanofibril and nanosilica. *Nanomaterials* **2021**, *11*, 2664. [CrossRef] [PubMed]
15. Yusefi, M.; Chan, H.Y.; Teow, S.Y.; Kia, P.; Soon, M.L.-K.; Sidik, N.A.B.C.; Shameli, K. 5-fluorouracil encapsulated chitosan-cellulose fiber bionanocomposites: Synthesis, characterization and in vitro analysis towards colorectal cancer cells. *Nanomaterials* **2021**, *11*, 1691. [CrossRef] [PubMed]
16. Pinto, R.J.B.; Lameirinhas, N.S.; Guedes, G.; da Silva, G.H.R.; Oskoei, P.; Spirk, S.; Oliveira, H.; Duarte, I.F.; Vilela, C.; Freire, C.S.R. Cellulose nanocrystals/chitosan-based nanosystems: Synthesis, characterization, and cellular uptake on breast cancer cells. *Nanomaterials* **2021**, *11*, 2057. [CrossRef] [PubMed]

nanomaterials

MDPI

Review

Recent Progress on the Characterization of Cellulose Nanomaterials by Nanoscale Infrared Spectroscopy

Qianqian Zhu [1], Rui Zhou [1], Jun Liu [1], Jianzhong Sun [1,*] and Qianqian Wang [1,2,3,4,*]

[1] Biofuels Institute, School of the Environment and Safety Engineering, Jiangsu University, Zhenjiang 212013, China; happyzqq@ujs.edu.cn (Q.Z.); zr2172773442@163.com (R.Z.); junliu115142@ujs.edu.cn (J.L.)

[2] Key Laboratory of Biomass Energy and Material, Institute of Chemical Industry of Forest Products, Chinese Academy of Forestry, Nanjing 210042, China

[3] State Key Laboratory of Bio-based Materials and Green Papermaking, Qilu University of Technology, Jinan 250353, China

[4] State Key Laboratory of New Textile Materials and Advanced Processing Technologies, Wuhan Textile University, Wuhan 430200, China

* Correspondence: jzsun1002@ujs.edu.cn (J.S.); wqq@ujs.edu.cn (Q.W.)

Abstract: Researches of cellulose nanomaterials have seen nearly exponential growth over the past several decades for versatile applications. The characterization of nanostructural arrangement and local chemical distribution is critical to understand their role when developing cellulose materials. However, with the development of current characterization methods, the simultaneous morphological and chemical characterization of cellulose materials at nanoscale resolution is still challenging. Two fundamentally different nanoscale infrared spectroscopic techniques, namely atomic force microscope based infrared spectroscopy (AFM-IR) and infrared scattering scanning near field optical microscopy (IR s-SNOM), have been established by the integration of AFM with IR spectroscopy to realize nanoscale spatially resolved imaging for both morphological and chemical information. This review aims to summarize and highlight the recent developments in the applications of current state-of-the-art nanoscale IR spectroscopy and imaging to cellulose materials. It briefly outlines the basic principles of AFM-IR and IR s-SNOM, as well as their advantages and limitations to characterize cellulose materials. The uses of AFM-IR and IR s-SNOM for the understanding and development of cellulose materials, including cellulose nanomaterials, cellulose nanocomposites, and plant cell walls, are extensively summarized and discussed. The prospects of future developments in cellulose materials characterization are provided in the final part.

Keywords: cellulose nanomaterials; cellulose nanocomposite; nanoscale resolution

check for **updates**

Citation: Zhu, Q.; Zhou, R.; Liu, J.; Sun, J.; Wang, Q. Recent Progress on the Characterization of Cellulose Nanomaterials by Nanoscale Infrared Spectroscopy. *Nanomaterials* **2021**, *11*, 1353. https://doi.org/10.3390/nano11051353

Academic Editor: Linda J. Johnston

Received: 14 April 2021
Accepted: 17 May 2021
Published: 20 May 2021

1. Introduction

Cellulose materials are renewable resources available in large quantities [1–3]. An in-depth understanding of the properties of cellulose materials will promote the development of renewable biomaterials and bioenergy, thereby enhancing the sustainability of our society [4,5]. In this review, cellulose materials are intended to contain a broader range of materials, including cellulose nanomaterials (CNMs), cellulose nanocomposites, and plant cell walls. CNMs are new classes of materials consisting of cellulose particles with at least one dimension smaller than 100 nm [6–11]. CNMs are known for their renewability, biocompatibility, biodegradability, low cytotoxicity, and excellent mechanical performances [12]. CNMs nanocomposites find many applications as packing materials, biomedical materials, smart materials, energy storage materials, etc. [13]. Due to all these advantages and potential applications, research on cellulose materials is under widespread scrutiny and investigation in both academic and industrial communities [14,15]. The excellent and unique properties of cellulose nanocomposites are determined by the distinct allocation and structural arrangement of cellulose particles within the composites, and also other factors such as the inherent properties of cellulose particles and matrix, and their concentration and interface. Understanding wood cell wall architecture provides vital information for improving the properties of wood composites [16]. Therefore, the consistency, reliability, accuracy, and spatial resolution of characterization in chemical distribution and structural arrangement are critical for increasing the mechanistic understanding of cellulose materials.

Best practices and techniques for cellulose materials characterization, such as CNMs, particularly in particle morphology, surface chemistry, mechanical properties, and toxicity, are extensively summarized and reviewed, which represents a solid foundation for CNMs characterization for both academic pursuits and industrial practice [6]. A collection of standardized protocols was also established or recommended. Cellulose material properties, such as particle morphology (scanning electron microscopy, SEM; transmission electron microscopy, TEM; atomic force microscopy, AFM), elemental composition (X-ray photoelectron spectroscopy, XPS; energy-dispersive X-ray spectroscopy, EDS; inductively coupled plasma mass spectrometry, ICP-MS), surface modification (Fourier transform infrared spectroscopy, FTIR; nuclear magnetic resonance, NMR), suspension properties (ultraviolet-visible spectroscopy, UV-Vis; zeta potential, rheology), solid-state properties (X-ray powder diffraction, XRD; Raman spectroscopy), and mechanical strength (dynamic mechanical analysis, DMA; tensile testing) were characterized by the aforementioned techniques in brackets. The need to characterize cellulose materials has driven significant advances in characterization capabilities.

The well-established tools, such as AFM and FTIR, were widely used for cellulose materials characterization. As a powerful and multifunctional technique, AFM offers a multitude of different acquisition modes and provides valuable information at the nanoscale on CNMs morphology [6,17], orientation [18,19], architecture deconstruction [20,21], mechanical strength [22], and adhesive properties [23]. FTIR spectroscopy is a bulk chemical structure analysis technique, which links IR absorptions with specific kinds of bonds, functional groups, and composite components. The area or volume-averaged spectral data cannot give regiospecific information. The powerful micro-FTIR spectroscopy enables the chemical mapping across the studied samples at the micron level [16,24–27]. The spatial resolution of micro FTIR spectroscopic imaging is determined by the diffraction limit associated with the mid-IR light at wavelengths of 2.5–10 μm. With the development of synchrotron-source IR spectroscopy and other approaches in recent years, it is now possible to provide an increase of the signal and better signal-to-noise ratio. However, the best achieved spatial resolution was still roughly limited to several micrometers. In brief, AFM with nanometer resolution can provide the morphology of cellulose materials without chemical information, whereas micro-FTIR provides chemical information with limited spatial resolution.

Although significant progress has been made with the aforementioned techniques for cellulose materials characterization, there are still many practical challenges that need to be efficiently and economically solved [4,28,29]. For example, the characterization of the interaction between CNMs and water molecules presents a practical challenge in dewatering, rehydration, and redispersing because of the strong tendency to form hydrogen bonding. Because of the nanoscale dimension, high surface area, and low percolation thresholds, the characterization of CNMs dispersibility in different media and their distribution in the composite matrix is challenging [6,10]. The interfaces (sharp two-dimensional boundaries between two phases) and interphases (transition zones between two phases in a composite) play a vital role in the mechanical properties of the resulting cellulose nanocomposites. Thus, understanding the nature of the interface/interphase is important when developing cellulose nanocomposites. Characterizing the interface and interphase of cellulose nanocomposites with morphological and chemical information simultaneously at nanoscale for a more fundamental understanding is a real challenge due to their small size and complex interactions, and also the lack of suitable tools. To meet these challenges, nanoscale IR spectroscopy including atomic force microscope based infrared spectroscopy (AFM-IR) and infrared scattering scanning near field optical microscopy (IR s-SNOM) have been adopted to realize nanoscale spatially resolved imaging for both morphological and chemical information by the integration of AFM with IR spectroscopy [30,31]. Nanoscale IR spectroscopy provides information-rich IR spectra that allow for nanoscale chemical identification of different groups and compositions based on standard FTIR libraries. The state-of-the-art AFM-IR and IR s-SNOM techniques can reach 10-nm spatial resolutions.

This review first describes the urgent need for consistent, reliable, and accurate characterization techniques with simultaneously morphologic and chemical information at the nanoscale for cellulose materials. As a potential solution, nanoscale IR spectroscopies, including AFM-IR and IR s-SNOM, are introduced in Section 2. Section 3 presents the state-of-the-art research applications where nanoscale IR spectroscopy leads as a novel multi-scale visualization and characterization technique for both morphological and chemical information of cellulose materials. Specific applications, including CNMs, CNMs nanocomposites, and plant cell walls, are emphasized. Section 4 summarizes the capabilities and limitations of nanoscale IR spectroscopy. Lastly, the prospects on the research direction by nanoscale IR spectroscopy for cellulose materials characterization are provided in Section 5.

2. Overview of AFM-IR and IR s-SNOM

Nanoscale infrared spectroscopies of both AFM-IR and IR s-SNOM are hybrid technologies that integrate the spatial resolution of AFM and the chemical analysis capability of IR spectroscopy [32]. With those nanoscale infrared spectroscopies, it is possible to determine and image local chemical composition below the diffraction limit. AFM-IR measures the photothermal-induced resonance effect by measuring either the local thermal expansion or temperature rise with AFM tip after IR absorption by a sample [33,34]. The thermal expansion or temperature change is proportional to the infrared light absorption coefficient of the sample. Thus, the detected signal can be converted into the vibrational spectrum of the sample. The IR s-SNOM technique, by contrast, determines the amount of scattered light from the specimen [35,36]. The light scattered was influenced by the comprehensive optical properties of the AFM probe, specimen, and underlying substrate. The schematic diagrams of AFM-IR, IR s-SNOM, and their differences in principles are shown in Figure 1. AFM-IR equipment consists of a pulsed tunable IR source with an AFM (Figure 1a,c). When the pulsed IR light is absorbed by the sample near the AFM tip, a rapid photothermal expansion of the sample occurs creating a transient cantilever oscillation, which can be further converted to a local IR absorption spectrum as a function of wavenumber [33]. The sharp AFM tip and infrared spectroscopy endow AFM-IR with the chemical analysis and imaging capabilities at the nanoscale. A pulsed tunable IR source with an AFM is also an essential accessory for IR s-SNOM (Figure 1b,d). On the other hand, IR s-SNOM collects the scattered near-field of light from a metallic AFM tip with the sample underneath [35]. The wavelength or wavenumber-dependent optical and chemical properties can be obtained via the collected light with amplitude and phase properties [36,37]. More basic principles on the theory for AFM-IR and IR s-SNOM techniques are available in [34,38–42].

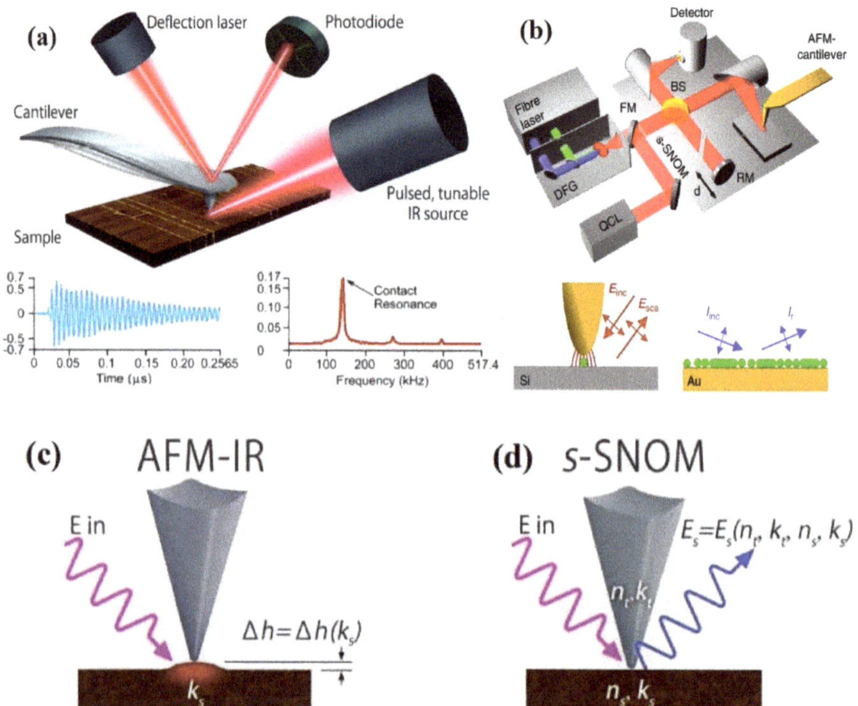

Figure 1. Schematic diagram of AFM-IR (**a**), IR s-SNOM (**b**), and their working principles (**c,d**). Reproduced with permission from [37,43]. Copyright Macmillan Publishers Limited, 2013; Copyright Microscopy Society of America; Copyright Anasys Instruments (Now Bruker Corporation), 2015.

As such, AFM-IR performs best on samples with a larger thermal expansion (the tendency of matter to change its volume and shape in response to temperature variation), whereas IR s-SNOM is the best technique for materials with a larger light scattering coefficient (the ability of matter to scatter photons). Both complementary modes can be acquired by combining AFM-IR and IR s-SNOM into a single instrument. This was achieved in nanoIR2-s™ (Anasys Instruments, Santa Barbara, CA, USA), neaSNOM system (Neaspec GmbH, Munich-Haar, Germany), and later versions by the pioneered companies of Anasys Instruments (Now part of Bruker Corporation) and Neaspec GmbH (Now Part of Attocube Systems), respectively.

3. AFM-IR and IR s-SNOM Application in Cellulose Materials

Advances of AFM-IR and IR s-SNOM technology make it possible to perform IR spectroscopic mapping and chemical analysis at the nanoscale with spatial resolution beyond the Abbe diffraction limit for the characterization of nanomaterials and nanostructures. AFM-IR is now recognized as one of the most important novel techniques for analyzing various systems, such as polymer blends, organic fibers, composites, multilayer thin films in addition to biological samples [33,34,44], while IR s-SNOM has demonstrated great popularity in studying monolayered and inorganic samples as well as soft materials [35,38,45].

In addition to its capability of achieving nanoscale chemical analysis and composition mapping, several practical benefits of AFM-IR and IR s-SNOM are listed below: (i) Simple sample preparation. Solution deposition, spin coating, and microtome are the most adopted processes for flat sample preparation. Top-down and top-side illumination endow a substantially broader range of samples with arbitrary thickness to be examined on arbitrary substrates [46]. (ii) High-speed and non-destructive measurement. Spectra can be obtained in seconds or less. (iii) Insensitivity to fluorescence and no need for staining steps. This is very useful, especially for characterizing the plant cell wall with lignin component. (iv) Rich, interpretable nanoscale infrared spectra. Nanoscale IR spectroscopy and imaging offer a wide range of morphological and chemical properties of the specimens at the nanoscale [31,40]. A preliminary comparison of the advantages and disadvantages of AFM-IR and IR s-SNOM is briefly summarized in Table 1.

Table 1. Advantages and disadvantages of AFM-IR and IR s-SNOM techniques.

Technique	Advantages	Disadvantages
AFM-IR	• Non-destructive, ambient conditions • Fine spatial resolution • morphological and chemical imaging • Suitable for organics, such as polymer, blends, and composites as well as biological samples • Point IR spectra acquisition • Directly correlates to FTIR	• Background interference • Tip contamination • Signal enhancement by gold-coated probes and substrates • Reduced peak intensity as compared to bulk IR spectra • Resolution and signal intensity is limited by sample properties (thickness, smoothness thermal diffusivity, etc.)
IR s-SNOM	• Non-destructive, ambient conditions • Fine spatial resolution • Optical properties including amplitude and phase • Suitable for inorganics, such as photonics and 2D materials that efficiently scatter light	• Background interference • Tip contamination • Reduced peak intensity as compared to bulk IR spectra • Need theoretical models to interpret data • Artifacts such as band distortion and thermal drift

Since its first demonstration to characterize cellulose materials, the application of nanoscale IR spectroscopy and imaging to study the properties of cellulose materials are receiving great attention from the research community. The progressive increase in the number of publications in recent years reflects the rising interest and importance of such new techniques for cellulose materials characterization. This paper aims to provide a thorough literature review on the characterization of cellulose materials using AFM-IR and IR s-SNOM. Table 2 summarizes the major applications of nanoscale IR spectroscopy techniques for cellulose materials.

Table 2. Application of AFM-IR and IR s-SNOM for the characterization of cellulose materials.

Sample	Sample Preparation	Instrument	AFM Tip	Spatial Resolution	Spectral Range /cm^{-1}	Research Area/Topic	Reference
Cellulose Nanomaterials							
Microfibrilated cellulose (MFC)	Solution deposition; spin coating; microtomed	NanoIR	NA.	50 nm	1200–1800	Local crystallinity analyses	[46]
Cellulose Nanocrystals (CNCs)	Single CNC particles	neaSNOM	Au-coated tips	nanoscale	800–3600	Cellulose polymorphy and sulfur analyses	[47]
Microcrystalline cellulose (MCC)	Solution deposition	NanoIR	NA.	NA.	1160, 1490, 1735	Distribution of functional groups	[48]
Single cotton cellulose fiber	Single fibers	NanoIR2	NA.	10 nm	2600–3800	Observation of water on cellulose surfaces	[49]
CNMs Nanocomposites							
CNCs/polyurethane (PU) foams	foam cross sections	neaSNOM system	Pt-Si or Au coated tips	a few ten nanometers	900–1400	CNCs distribution in PF foam	[50]
Nanocellulose fibrils with high lignin content (NCFHL)/polylactic acid (PLA) composites	Microtomed; several nanometers	NanoIR2	Au coated silicon nitride tip	tens of nanometers	800–4000	Dispersion of various phases and interfacial regions	[51]
CNCs/Polyacrylonitrile (PAN) nanofiber	Nanofiber from electrospinning	VistaScope microscope + quantum cascade lasers	NCH-Au 300 kHz noncontact tip	nanoscale resolution	800–1800	Heterogeneous substructure of PAN/CNCs nanofiber	[52]
CNCs/ poly(butylene adipate-co-terephthalate) (PBAT) nanocomposites	Suspension deposition	NanoIR2-s	AFM tip	NA.	1600–1800	Dispersion of CNCs and modified CNCs in PBAT	[53]
Starch/cellulose nanofibers (CNF) nanocomposites	Film casting on gold-coated silicon substrate	NanoIR2-s	AFM tip	10 nm	1530–1845	Topography correlated with local chemical analyses	[54,55]

Table 2. *Cont.*

Sample	Sample Preparation	Instrument	AFM Tip	Spatial Resolution	Spectral Range /cm^{-1}	Research Area/Topic	Reference
Plant Cell Wall							
Wood cell wall	Microtomed; 500 nm	NanoIR	AFM tip	100 nm	1200–1800	Microdomains spectroscopic characterization	[56]
Wood cell wall	Microtomed; 5 μm	NanoIR2	Au-coated tip	nanoscale resolution	1530–1810	Chemical alterations and inhomogeneity of cell wall	[57,58]
Wood cell wall	Microtomed; 10 μm	NanoIR	AFM tip	NA.	1200–1800	Heartwood formation process	[59]
Wood cell wall	Microtomed; smooth surfaces	Homemade nanoscale IR	Pt-coated silicon probe	16 nm	900–1800	Nanoscale chemical features of wood substrates	[60]
Wood cell wall	Microtomed	NanoIR2	NA.	sub-100 nm	1550–1800	Local chemical changes and lignin rearrangement	[61]
Wood cell wall /polymer composites	Microtomed; 200 nm	NanoIR2	Au-coated silicon nitride tip	100 nm	900–1800	Molecular-scale interactions of polymer and cell wall	[62,63]

NA.: Not Available. The information is not provided in the cited reference.

3.1. AFM-IR and IR s-SNOM Application in CNMs Characterization

Elucidation of the structural organization of CMMs and chemical modified cellulose materials is crucial to understand their role in different applications. The nanoscale IR spectroscopy provides structure arrangement and chemical composition information from nanoscale domains and structural organization of CNMs. The pioneering AFM-IR experiments that characterize microfibrillated cellulose (MFC) as a paper additive were conducted by Marcott et al. [46]. MFC is short rod-like cellulose fibrils with high crystallinity. As the first example for cellulose materials characterized by AFM-IR, the AFM topography image (Figure 2a) and IR absorbance image (Figure 2b) acquired with the light source tuned to 1360 cm^{-1} have been demonstrated as nanoscale chemical spectroscopy with a spatial resolution of 50 nm. The high crystalline domains of MFC in the paper were visualized.

Nanoscale IR spectroscopy can effectively characterize the distribution of functional groups of cellulose nanomaterials and cellulose derivatives at the nanoscale, such as sulfate half-ester groups and aromatic groups [47,48]. The structurally and chemically different CNC particles obtained from sulfuric acid hydrolysis (AcCNC) and enzymatic hydrolysis (EnCNC) were examined with a neaSNOM system using Au-coated AFM tips as shown in Figure 2d–g. The peaks at 814 cm^{-1} and 1250 cm^{-1} attributed to sulfate groups indicated the esterification of surface hydroxy groups of cellulose. Styrene grafting, chloroacetylation, amination, and protonation of MCC were also verified by tuning the IR source at the characteristic wavelengths of 1160, 1490, and 1735 cm^{-1}. The AFM-IR result demonstrated that grafting and copolymerization of styrene occurred mainly on the cellulose surface.

As compared to bulk IR spectroscopy, nanoscale IR spectroscopy has great potential for an in-depth analysis of the mechanism of interactions between CNMs and small molecules, such as water [49]. Figure 3a shows slight differences of the OH-stretch band in the 3800–2600 cm^{-1} region in the conventional ART-IR spectra between different dried cotton cellulose. Due to the low resolution and high noise level, bulk ART-IR cannot provide genuine microscopic information on the bound water localized at the outermost surface of differently dried cotton cellulose. This limitation of bulk ART-IR was overcome by AFM-IR. The difference in the coefficient of thermal expansion between water and cellulose made the AFM-IR measurement possible. As a surface-sensitive approach, AFM-IR indicated how and where the bound water exists on the cellulose. The state of hydrogen bonding in bound water is different from that in bulk water. The bulk water has a broad trapezoidal shape spectrum, whereas the bound water on the cellulose surface exhibits two decoupled stretching modes of OH groups, as shown in Figure 3b, which originate from the effects of the hydrophobic air-water interface (at lower wavenumber side) and hydrophilic water–cellulose interface (at higher wavenumber side), respectively.

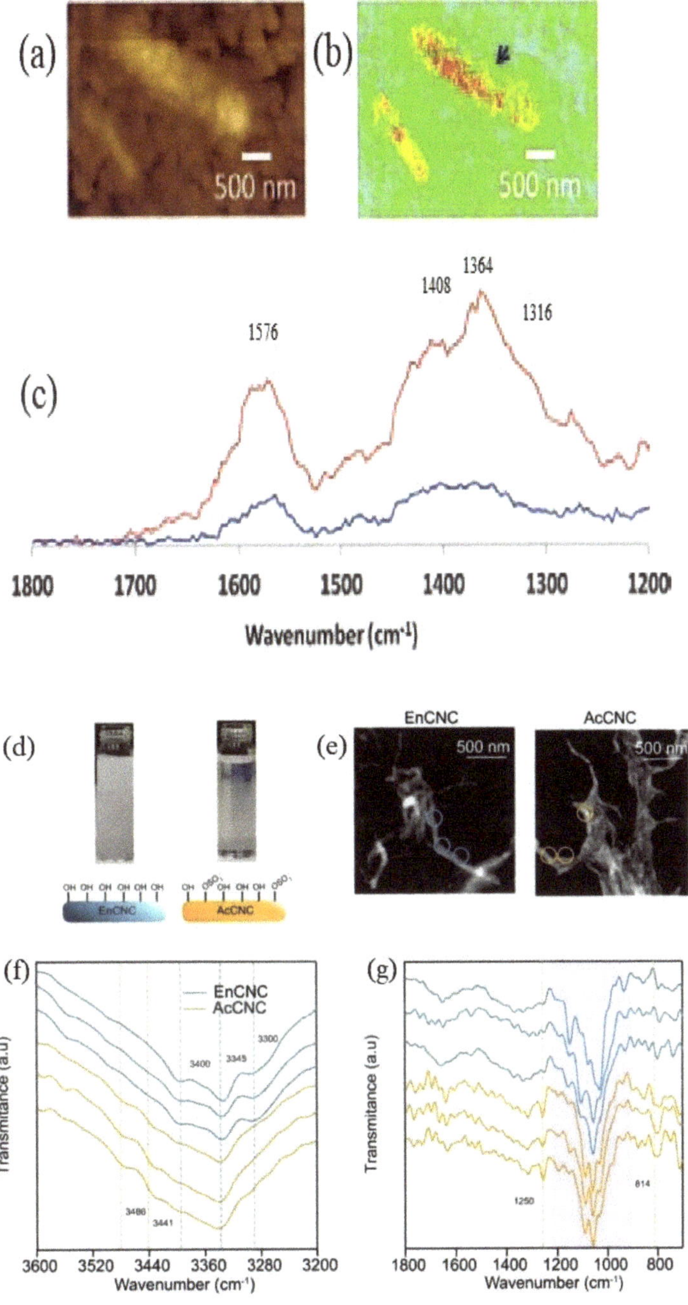

Figure 2. (**a**) AFM topography image of MFC in cellulose matrix, (**b**) NanoIR image at 1360 cm^{-1}, (**c**) NanoIR spectra; (**d**) CNCs suspension produced by enzymatic and acid hydrolysis, (**e**) Topography images from EnCNC and AcCNC, (**f**) NanoIR spectra for EnCNC and AcCNC nanocrystals in 3600 to 3200 cm^{-1} (**f**) and 1800 to 750 cm^{-1} (**g**). Reproduced with permission from [46]. Copyright Cambridge University Press, 2012. Reproduced from [47].

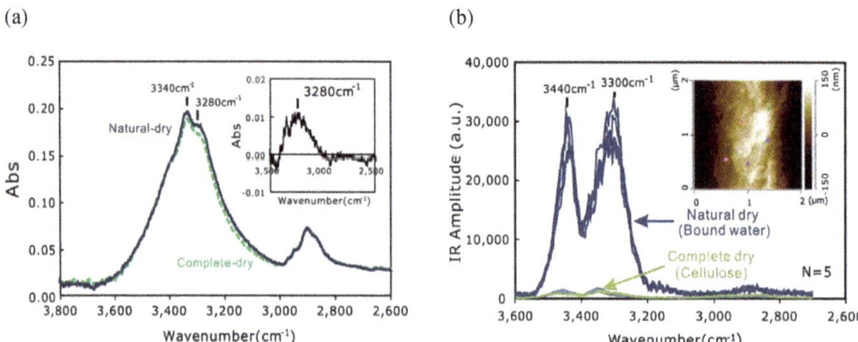

Figure 3. (**a**) Conventional ATR-IR spectra of dried cotton cellulose; (**b**) Direct observation of the interaction between water and cellulose by AFM-IR. Reproduced with permission from [49]. Copyright American Chemical Society, 2020.

3.2. AFM-IR and IR s-SNOM Application in CNMs Nanocomposites Characterization

CNMs nanocomposites are becoming increasingly important in the academic and industrial community, where CNMs with enhanced properties are being added to bulk polymers to achieve improved properties and performances [43]. Well dispersed cellulose nanomaterials can enhance the properties of cellulose nanocomposites. Nonetheless, little is understood about the exact mechanism and extent of interfacial chemistry in the interphase regions. This requires a localized probe to fully understand the local structure and chemistry, interphase formation, interfacial chemistry, and mechanical properties of CNMs nanocomposites. To achieve a deep understanding of the micro/nano arrangements and structures of CNCs, the distribution of CNCs in polyurethane foam nanocomposites was verified by IR s-SNOM imaging and nano-FTIR spectroscopy with a neaSNOM system [50]. Microdomains of strong IR phase contrast at 1050 cm^{-1} indicated the even distribution of CNCs in the nanocomposites. Similar research was conducted to study the distribution of nanocellulose fibrils with high lignin content (NCFHL) in the PLA matrix and their interface properties [51]. Ultrathin cellulose nanocomposite crosssection was used for AFM-IR imaging on the silicon substrate. NanoIR spectra complement well with the conventional FTIR spectra as shown in Figure 4. Slight differences in peak position and intensity were detected, which may be more local characteristics reflected by nanoIR spectra. The intense IR absorption indicates the presence of higher concentrations of nanocellulose fibrils (Figure 4c). Nanoscaled IR maps were converted into binary images to examine the interphase distribution. The quantitative analysis of the binary images indicated that approximately 22% of the total area was filled with NCFHL fibrils, while neat PLA occupied about 32% as a separate phase. Approximately 46% region in the binary map was attributed to the transition phase or interfacial area. This large proportion of interfacial area within the specimen was ascribed to the reaction between highly dispersed NCFHL and PLA molecules, and also lignin as a strong compatibilizer during the pressing of the biocomposites at high temperature.

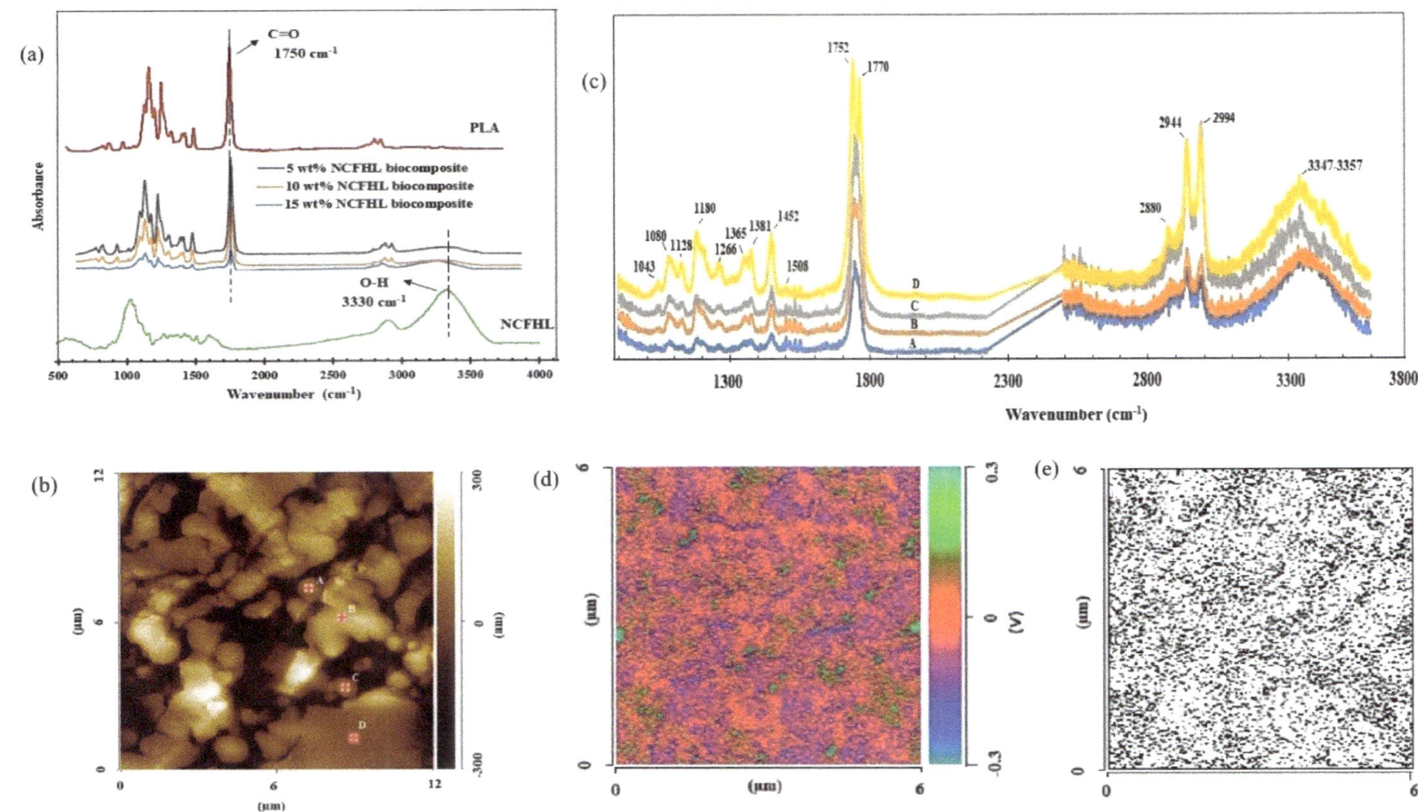

Figure 4. (**a**) Bulk FTIR spectra for nanocellulose fibrils, neat PLA, and their nanocomposites (**b**) AFM topography image of the nanocomposites (**c**). Corresponding NanoIR spectra obtained for areas in (**b**). (**d**) NanoIR (Anasys Instruments, Santa Barbara, CA, USA) images with the IR laser at 3330 cm⁻¹ (nanocellulose fibrils characteristic peak), and its corresponding binary image (**e**). Reproduced with permission from [51]. Copyright American Chemical Society, 2018.

Local and regional distribution of CNF and starch distribution was correlated to the topographic information with the local chemical analysis [54,55]. A new peak at a wavelength of 1630 cm^{-1} in the local IR spectrum indicated the formation of hydrogen bonding between cellulose and starch. The dispersion behavior of adipic acid molecules modified CNCs in polybutylene adipate-co-terephthalate (PBAT) was examined by AFM-IR at 1700 cm^{-1} for C=O absorption [53]. The good dispersion and interaction of adipic acid-modified CNCs in PBAT were attributed to the hydrophilic core and hydrophobic shell structure of modified CNCs.

Tip functionalization could be utilized to improve the spatial resolution of nanoscale IR spectroscopy [52]. The refractive index contrast of CNMs was greatly enhanced by using a polydimethylsiloxane (PDMS) functionalized tip. The fine structure of CNCs alignment inside the PAN-CNCs fiber can be demonstrated. The strands of polyacrylonitrile-cellulose nanocrystals (PAN-CNCs) nanofibers show strong thermal expansion at 1462 cm^{-1}, whereas the bare CNCs debris exhibits weak thermal expansion. In contrast, the bare CNCs particles illustrate a strong thermal expansion at 1027 cm^{-1}, while the CNCs inside the PAN fiber display weak signals. The refractive indices for PAN and CNCs are comparable to each other. The refractive index contrast between CNCs and PAN was greatly enhanced with the assistance of a polydimethylsiloxane (PDMS) functionalized tip. With the enhanced refractive index contrast, the substructure of PAN-CNCs nanofiber and bare CNCs particles were simultaneously visualized. The CNCs alignment within the PAN-CNCs nanofibers was clearly manifested. Other functionalized tips such as Pt or Au coated AFM tips were also used to enhance the signals [47,50,51].

3.3. AFM-IR and IR s-SNOM Application in CNMs Nanocomposites Characterization

It is of great interest to characterize the nanostructure and local arrangement of plant cell wall components [21]. The local properties of plant cell walls were strongly correlated with the nanoscale architecture of its components. The morphological and chemical structure of wood cell walls at the nanoscale resolution or even molecular level could promote a deep understanding of fine structures of plant cell walls after pretreatment and modification. AFM-IR provides a new method to investigate the chemistry of plant cell walls at the nanoscale resolution without employing special treatments.

To our best knowledge, the first study using nanoscale IR spectroscopy to characterize the wood cell wall was reported by Marcott et al. using an AFM-IR instrument at spatial resolutions on the order of 100 nm with a tunable IR laser source that illuminates the sample from below [56]. The underneath laser illumination requires that the wood samples must be thin enough. The optimal thickness is less than 1 μm (500 nm, in this study). IR spectra as a function of spatial position for both transverse and longitudinal cross-sections of the untreated and acetylated wood samples were examined by AFM-IR, as shown in Figure 5. Lignin content in compound corner middle lamella as reflected by the band near 1500 cm^{-1} was higher than that in the S2-layer [56]. The acetylation intensified the carbonyl band at 1736 cm^{-1}. The broadband IR spectrum obtained at any arbitrary point provides insight into microdomain composition and leads to an increased understanding of the cell wall structure and properties of biomaterials. Updated versions of AFM-IR instruments with top-down or top-side illumination do not have such limitations in that specimens of arbitrary thickness can be examined on arbitrary substrates. Similarly, the distribution of functional groups in native and chemical modified wood was examined with an excellent spatial resolution of 16 nm [60].

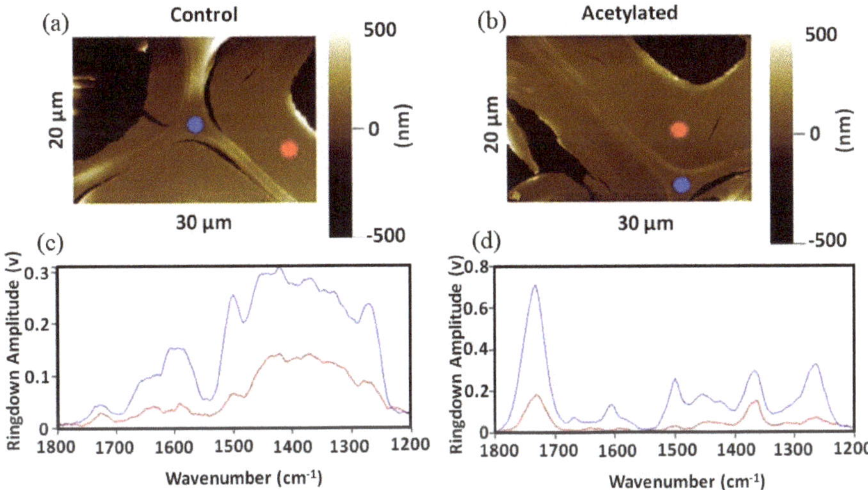

Figure 5. AFM topography images of (**a**) a control wood sample and (**b**) an acetylated wood sample, and their corresponding nanoIR spectra (**c,d**) at indicated locations. Reproduced with permission from [56]. Copyright Anasys Instruments (Now Bruker Corporation), 2012.

NanoIR imaging has also been employed to capture the removal of lignin occurring upon pretreatments of the plant cell wall [58]. Local IR spectra of the pretreated sample exhibited a significant drop in lignin content, confirming alterations in topography. The correlation of the AFM morphological image, IR-sensitive images, the mechanical phase image enabled a deep understanding of mechanical properties together with the chemical components and structure of cell walls [64]. The nanoscale compositional variations in plant cell walls characterized by nanoscale IR spectroscopy were also used to interpret plant cell wall physiological phenomena, such as heartwood formation and water transport in the xylem [59,65]. The spectral variations at 1660 cm^{-1} and 1640 cm^{-1} in heartwood and sapwood were identified. The intensity of the peak at 1660 cm^{-1}, which was assigned to the ethylenic C=C and C=O bond stretches of lignin, decreased from sapwood to heartwood, whereas the intensity of the peak at 1660 cm^{-1} attributed to a carbonyl stretching and hydrogen bonding to the carbonyl group increased. The possible explanation for this variation is that phenolic precursors of extractives accumulate in the sapwood, are oxidized and condensed in the transition zone, and diffuses into the heartwood. Distributions of hydrophobic compounds and compounds with a high electric potential of pit membranes in water-conducting cells may influence the water transport system in plants. Further research is required to illustrate the effects of plant species and water availability variations.

The accurate evaluation of the interphase properties is critical to understanding how the polymer interacts with the wood cell wall components and optimizing the design and fabrication of the wood-plastic composites (WPCs). The specific molecular-scale penetration and interactions between phenol-formaldehyde (PF) resin and wood cell wall were in situ identified by AFM-IR for the first time [62]. The intensity of the characteristic peaks for cell wall and PF resin gradually changes across the distance from the resin region, resin-cell wall interphase area, to cell wall area as shown in the AFM-IR images. Similarly, the penetration of isocyanates (pMDI) in the wood cell wall was also evaluated by AFM-IR [63]. The molecular-scale penetration and interactions between pMDI and cell wall strengthened the connections between the cell wall and polymer and thus improved the mechanical strength.

3.4. AFM-IR and IR s-SNOM Application for Other Cellulose-Containing or Cellulose-Derived Materials

It was challenging to characterize the drug–polymer miscibility due to the similarity in the glass transition temperature and small size domain. Recently, AFM-IR has been applied to evaluate the miscibility of drug and hydroxypropyl methylcellulose as shown in Figure 6 [66,67]. The AFM-IR data showed that the phase separation in a drug (Itraconazole, Telaprevir) and hydroxypropyl methylcellulose dispersion occurred at the submicron scale. These results provide new insights into the interphase/interface behavior of drug–polymer dispersions.

Atomic force microscopy-nanoscale infrared images of a 30:70
itraconazole (ITZ)- hydroxypropylmethylcellulose (HPMC) film

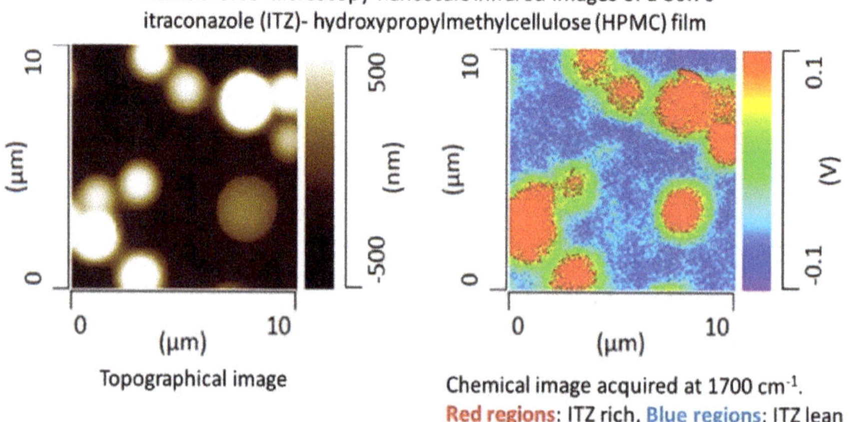

Figure 6. The miscibility of itraconazole-hydroxypropyl methylcellulose blends (30:70) characterized by AFM-IR. Reproduced with permission from reference [66]. Copyright American Chemical Society, 2015.

Cellulose fiber was used as a precursor fiber for carbon fiber. The evolution of functional groups (C–O and C=O) and the changes in micro-domain structure during the synthesis of cellulose-based carbon fibers were examined by nanoscale infrared spectroscopy [68]. The IR spectra of C–O and C=O indicated that the skin and core of carbon fibers were not pyrolyzed homogenously. AFM-IR techniques have also been used to detect the polypyrrole wearable nanofilm devices on the surface of cellulose-based substrates [69].

4. Challenge and Limitation of Nanoscale IR Spectroscopy in Cellulose Materials Characterization

Although offering remarkable advantages, nanoscale IR spectroscopy and imaging also have some limitations in cellulose materials characterization [33,34]. The resolution of nanoscale IR spectroscopy is often limited by the properties of cellulose specimens, such as thickness, roughness, and thermal diffusivity. What's more, these influences are wavelength related leading to more complex changes in relative signal intensities. Better resolution of experimental data may result in poor reproducibility. This poor reproducibility may not only be due to the differences in sample preparation but also closely related to the chemical inhomogeneity of the cellulose materials. This is especially true for plant cell walls. Thickness-induced chemical shifts of IR s-SNOM resulted in chemical and structural analysis more complex than those for conventional bulk spectra. The versatility of nanoscale IR spectroscopy and imaging is at the cost of the ability to directly correlate the obtained spectra information to the quantitative characterization of chemical and structural properties [70,71]. Complex data processing methods with certain degrees of signal degradation still cannot eliminate background interferences. A huge challenge for AFM-IR is the difficulty to quantify the relationship between the AFM-IR tip-sample contact dynamics and signal intensity, which depends on the stability of the AFM operation and the sample local thermomechanical properties [72]. To date, the ability of AFM-IR as an analytical tool to characterize the single cellulose molecule and single chemical bond scale has remained elusive over a wide range of cellulose materials. The limited resolution was due to a relatively large volume excited by IR lasers in the sample. The application of AFM-IR was limited to relatively flat and large self-assembled aggregates or monolayers composed of several hundreds of molecules. Many spatial resolutions during nanoscale IR measurements are not determined. This may be due to the difficulty in determining the spatial resolution of thick cellulose samples with irregular geometry and unclear margin. In general, IR spectroscopy is sensitive to water molecular and atmospheric gases. This is also true for nanoscale IR spectroscopy. Cellulose, as a hydrophilic and hygroscopic material, is also sensitive to water. There is no doubt that humidity will have a great influence on the characterization of cellulose materials by Nanoscale IR spectroscopy with such high resolution. Unfortunately, the effects of humidity on the nanoscale IR spectroscopy measurements of cellulose materials were not systematically studied. In some measurements, humidity data were not even provided. Recently, nanoscale IR spectroscopy instruments with humidity and temperature control accessories are available, which may be able to eliminate the interference of humidity and temperature on the characterization of cellulose materials.

5. Summary and Outlook

Nanoscale IR spectroscopy provides insightful data on the chemical structure of cellulose materials with nanoscale spatial resolution. The most recent advances of AFM-IR and IR s-SNOM on the characterization of CNMs, cellulose nanocomposites, and plant cell walls were extensively explored. Progress is still underway, and nanoscale IR spectroscopy now offers an exciting future for the application of cellulose materials. AFM-IR and IR s-SNOM have been used to identify and quantify polymer components in blends, characterize interfaces/interphase in composites, assess the local crystallization, and even reverse engineer multilayer films. Additionally, morphological and chemical maps could be further processed to get rich information. These techniques and protocols can also be applied to characterize cellulose materials. More than techniques for compositional analysis and chemical mapping at the nanoscale spatial resolution as discussed above, AFM-IR and IR s-SNOM can also obtain mechanical, thermal, and electrical property mapping at employed wavelength or spectra at employed positions for cellulose nanocomposites. Nevertheless, there are still many other opportunities to further anticipated developments. With the rapid theoretical and technical improvement in nanoscale IR spectroscopy, it is expected that the characterization of cellulose materials by AFM-IR and IR s-SNOM will increase significantly in the coming decades. In the foreseeable future, a more mechanistic understanding of CNMs composites will be obtained and more powerful CNMs nanocomposites will be developed.

Author Contributions: Conceptualization, Q.Z., and Q.W.; resources, J.S.; writing—original draft preparation, Q.Z. and Q.W.; writing—review and editing, Q.Z., R.Z., J.L., J.S. and Q.W.; project administration, J.S. and Q.W; funding acquisition, Q.Z., J.S., and Q.W. All authors have read and agreed to the published version of the manuscript.

Funding: The authors would like to thank the open funds from the Key Laboratory of Biomass Energy and Material of Jiangsu Province (JSBEM202012), State Key Laboratory of Biobased Material and Green Papermaking (GZKF202035), and State Key Laboratory of New Textile Materials and Advanced Processing Technologies (FZ2020018). This work is also funded by the National Natural Science Foundation of China (31300493), National Key R&D Program of China (2018YFE0107100), PAPD, and fund for Self-made Analytical Instruments of Jiangsu University (5623080004).

Data Availability Statement: Not applicable.

Conflicts of Interest: The authors declare no competing financial interest.

References

1. Klemm, D.; Heublein, B.; Fink, H.-P.; Bohn, A. Cellulose: Fascinating Biopolymer and Sustainable Raw Material. *Angew. Chem. Int. Ed.* **2005**, *44*, 3358–3393. [CrossRef]
2. Klemm, D.; Kramer, F.; Moritz, S.; Lindström, T.; Ankerfors, M.; Gray, D.; Dorris, A. Nanocelluloses: A New Family of Nature-Based Materials. *Angew. Chem. Int. Ed.* **2011**, *50*, 5438–5466. [CrossRef] [PubMed]
3. Wang, Q.; Sun, J.; Yao, Q.; Ji, C.; Liu, J.; Zhu, Q. 3D printing with cellulose materials. *Cellulose* **2018**, *25*, 4275–4301. [CrossRef]
4. Ray, U.; Zhu, S.; Pang, Z.; Li, T. Mechanics Design in Cellulose-Enabled High-Performance Functional Materials. *Adv. Mater.* **2020**, 2002504. [CrossRef] [PubMed]
5. Li, T.; Chen, C.; Brozena, A.H.; Zhu, J.Y.; Xu, L.; Driemeier, C.; Dai, J.; Rojas, O.J.; Isogai, A.; Wågberg, L.; et al. Developing fibrillated cellulose as a sustainable technological material. *Nat. Cell Biol.* **2021**, *590*, 47–56. [CrossRef]
6. Foster, E.J.; Moon, R.J.; Agarwal, U.P.; Bortner, M.J.; Bras, J.; Camarero-Espinosa, S.; Chan, K.J.; Clift, M.J.D.; Cranston, E.D.; Eichhorn, S.J.; et al. Current characterization methods for cellulose nanomaterials. *Chem. Soc. Rev.* **2018**, *47*, 2609–2679. [CrossRef] [PubMed]
7. Moon, R.J.; Martini, A.; Nairn, J.; Simonsen, J.; Youngblood, J. Cellulose nanomaterials review: Structure, properties and nanocomposites. *Chem. Soc. Rev.* **2011**, *40*, 3941–3994. [CrossRef] [PubMed]
8. Zhu, Q.; Yao, Q.; Sun, J.; Chen, H.; Xu, W.; Liu, J.; Wang, Q. Stimuli induced cellulose nanomaterials alignment and its emerging applications: A review. *Carbohydr. Polym.* **2020**, *230*, 115609. [CrossRef] [PubMed]
9. Zhu, Q.; Liu, S.; Sun, J.; Liu, J.; Kirubaharan, C.J.; Chen, H.; Xu, W.; Wang, Q. Stimuli-responsive cellulose nanomaterials for smart applications. *Carbohydr. Polym.* **2020**, *235*, 115933. [CrossRef]
10. Wang, Q.; Yao, Q.; Liu, J.; Sun, J.; Zhu, Q.; Chen, H. Processing nanocellulose to bulk materials: A review. *Cellulose* **2019**, *26*, 7585–7617. [CrossRef]
11. Vanderfleet, O.M.; Cranston, E.D. Production routes to tailor the performance of cellulose nanocrystals. *Nat. Rev. Mater.* **2021**, *6*, 124–144. [CrossRef]
12. Clarkson, C.M.; Azrak, S.M.E.A.; Forti, E.S.; Schueneman, G.T.; Moon, R.J.; Youngblood, J.P. Recent Developments in Cellulose Nanomaterial Composites. *Adv. Mater.* **2020**, 2000718. [CrossRef] [PubMed]
13. Mu, R.; Hong, X.; Ni, Y.; Li, Y.; Pang, J.; Wang, Q.; Xiao, J.; Zheng, Y. Recent trends and applications of cellulose nanocrystals in food industry. *Trends Food Sci. Technol.* **2019**, *93*, 136–144. [CrossRef]
14. Mishnaevsky, L.; Mikkelsen, L.P.; Gaduan, A.N.; Lee, K.-Y.; Madsen, B. Nanocellulose reinforced polymer composites: Computational analysis of structure-mechanical properties relationships. *Compos. Struct.* **2019**, *224*, 111024. [CrossRef]
15. Jabbar, A.; Militký, J.; Wiener, J.; Kale, B.M.; Ali, U.; Rwawiire, S. Nanocellulose coated woven jute/green epoxy composites: Characterization of mechanical and dynamic mechanical behavior. *Compos. Struct.* **2017**, *161*, 340–349. [CrossRef]
16. Zhao, Y.; Man, Y.; Wen, J.; Guo, Y.; Lin, J. Advances in Imaging Plant Cell Walls. *Trends Plant Sci.* **2019**, *24*, 867–878. [CrossRef]
17. Usov, I.; Nyström, G.; Adamcik, J.; Handschin, S.; Schütz, C.; Fall, A.; Bergström, L.; Mezzenga, R. Understanding nanocellulose chirality and structure–properties relationship at the single fibril level. *Nat. Commun.* **2015**, *6*, 7564. [CrossRef]
18. Gray, D.G.; Mu, X. Chiral Nematic Structure of Cellulose Nanocrystal Suspensions and Films; Polarized Light and Atomic Force Microscopy. *Materials* **2015**, *8*, 7873–7888. [CrossRef]
19. Skogberg, A.; Mäki, A.-J.; Mettänen, M.; Lahtinen, P.; Kallio, P. Cellulose Nanofiber Alignment Using Evaporation-Induced Droplet-Casting, and Cell Alignment on Aligned Nanocellulose Surfaces. *Biomacromolecules* **2017**, *18*, 3936–3953. [CrossRef]
20. Chundawat, S.P.S.; Donohoe, B.S.; Sousa, L.D.C.; Elder, T.; Agarwal, U.P.; Lu, F.; Ralph, J.; Himmel, M.E.; Balan, V.; Dale, B.E. Multi-scale visualization and characterization of lignocellulosic plant cell wall deconstruction during thermochemical pretreatment. *Energy Environ. Sci.* **2011**, *4*, 973–984. [CrossRef]
21. Ding, S.-Y.; Liu, Y.-S.; Zeng, Y.; Himmel, M.E.; Baker, J.O.; Bayer, E.A. How Does Plant Cell Wall Nanoscale Architecture Correlate with Enzymatic Digestibility? *Science* **2012**, *338*, 1055–1060. [CrossRef] [PubMed]
22. Iwamoto, S.; Kai, W.; Isogai, A.; Iwata, T. Elastic Modulus of Single Cellulose Microfibrils from Tunicate Measured by Atomic Force Microscopy. *Biomacromolecules* **2009**, *10*, 2571–2576. [CrossRef] [PubMed]
23. Lahiji, R.R.; Xu, X.; Reifenberger, R.; Raman, A.; Rudie, A.; Moon, R.J. Atomic Force Microscopy Characterization of Cellulose Nanocrystals. *Langmuir* **2010**, *26*, 4480–4488. [CrossRef]
24. Xiao, N.; Bock, P.; Antreich, S.J.; Staedler, Y.M.; Schönenberger, J.; Gierlinger, N. From the Soft to the Hard: Changes in Microchemistry During Cell Wall Maturation of Walnut Shells. *Front. Plant Sci.* **2020**, *11*, 466. [CrossRef] [PubMed]
25. Labb, N.; Rials, T.G.; Kelley, S.S. FTIR Imaging of Wood and Wood Composites. In *Characterization of the Cellulosic Cell Wall*; Stokke, D.D., Groom, L.H., Eds.; Wiley: Hoboken, NJ, USA, 2008; pp. 110–122.
26. Mukherjee, T.; Tobin, M.J.; Puskar, L.; Sani, M.-A.; Kao, N.; Gupta, R.K.; Pannirselvam, M.; Quazi, N.; Bhattacharya, S. Chemically imaging the interaction of acetylated nanocrystalline cellulose (NCC) with a polylactic acid (PLA) polymer matrix. *Cellulose* **2017**, *24*, 1717–1729. [CrossRef]
27. Guo, X.; Wu, Y. IN SITU visualization of water adsorption in cellulose nanofiber film with micrometer spatial resolution using micro-FTIR imaging. *J. Wood Chem. Technol.* **2018**, *38*, 361–370. [CrossRef]
28. Yang, X.; Biswas, S.K.; Han, J.; Tanpichai, S.; Li, M.; Chen, C.; Zhu, S.; Das, A.K.; Yano, H. Surface and Interface Engineering for Nanocellulosic Advanced Materials. *Adv. Mater.* **2020**, 2002264. [CrossRef]

29. Keplinger, T.; Wittel, F.K.; Rüggeberg, M.; Burgert, I. Wood Derived Cellulose Scaffolds—Processing and Mechanics. *Adv. Mater.* **2020**, 2001375. [CrossRef] [PubMed]
30. Dazzi, A.; Prazeres, R.; Glotin, F.; Ortega, J.M. Local infrared microspectroscopy with subwavelength spatial resolution with an atomic force microscope tip used as a photothermal sensor. *Opt. Lett.* **2005**, *30*, 2388–2390. [CrossRef] [PubMed]
31. Dendisová, M.; Jeništová, A.; Parchaňská-Kokaislová, A.; Matějka, P.; Prokopec, V.; Švecová, M. The use of infrared spectroscopic techniques to characterize nanomaterials and nanostructures: A review. *Anal. Chim. Acta* **2018**, *1031*, 1–14. [CrossRef] [PubMed]
32. Unger, M.; Marcott, C. Recent Advances and Applications of Nanoscale Infrared Spectroscopy and Imaging. In *Encyclopedia of Analytical Chemistry*; Wiley: Hoboken, NJ, USA, 2017; pp. 1–26.
33. Dazzi, A.; Prater, C.B. AFM-IR: Technology and Applications in Nanoscale Infrared Spectroscopy and Chemical Imaging. *Chem. Rev.* **2017**, *117*, 5146–5173. [CrossRef] [PubMed]
34. Kurouski, D.; Dazzi, A.; Zenobi, R.; Centrone, A. Infrared and Raman chemical imaging and spectroscopy at the nanoscale. *Chem. Soc. Rev.* **2020**, *49*, 3315–3347. [CrossRef] [PubMed]
35. Rao, V.J.; Matthiesen, M.; Goetz, K.P.; Huck, C.; Yim, C.; Siris, R.; Han, J.; Hahn, S.; Bunz, U.H.F.; Dreuw, A.; et al. AFM-IR and IR-SNOM for the Characterization of Small Molecule Organic Semiconductors. *J. Phys. Chem. C* **2020**, *124*, 5331–5344. [CrossRef]
36. Meyns, M.; Primpke, S.; Gerdts, G. Library Based Identification and Characterisation of Polymers with Nano-Ftir and Ir-Ssnom Imaging. *Anal. Methods* **2019**, *11*, 5195–5202. [CrossRef]
37. Amenabar, I.; Poly, S.; Nuansing, W.; Hubrich, E.H.; Govyadinov, A.A.; Huth, F.; Krutokhvostov, R.; Zhang, L.; Knez, M.; Heberle, J.; et al. Structural analysis and mapping of individual protein complexes by infrared nanospectroscopy. *Nat. Commun.* **2013**, *4*, 2890. [CrossRef] [PubMed]
38. Jarzembski, A.; Shaskey, C.; Park, K. Review: Tip-based vibrational spectroscopy for nanoscale analysis of emerging energy materials. *Front. Energy* **2018**, *12*, 43–71. [CrossRef]
39. Prater, C.; Kjoller, K.; Shetty, R. Nanoscale infrared spectroscopy. *Mater. Today* **2010**, *13*, 56–60. [CrossRef]
40. Kenkel, S.; Mittal, S.; Bhargava, R. Closed-loop atomic force microscopy-infrared spectroscopic imaging for nanoscale molecular characterization. *Nat. Commun.* **2020**, *11*, 1–10. [CrossRef]
41. Huth, F.; Schnell, M.; Wittborn, J.; Ocelic, N.; Hillenbrand, R. Infrared-spectroscopic nanoimaging with a thermal source. *Nat. Mater.* **2011**, *10*, 352–356. [CrossRef]
42. Huth, F. Nano-Ftir—Nanoscale Infrared Near-Field Spectroscopy. Ph.D. Thesis, Euskal Herriko Unibertsitatea-Universidad del Pais Vasco, Bilbao, Spain, 2015.
43. Marcott, C.; Lo, M.; Dillon, E.; Kjoller, K.; Prater, C. Interface Analysis of Composites Using AFM-Based Nanoscale IR and Mechanical Spectroscopy. *Microsc. Today* **2015**, *23*, 38–45. [CrossRef]
44. Nguyen-Tri, P.; Ghassemi, P.; Carriere, P.; Nanda, S.; Assadi, A.A.; Nguyen, D.D. Recent Applications of Advanced Atomic Force Microscopy in Polymer Science: A Review. *Polymers* **2020**, *12*, 1142. [CrossRef] [PubMed]
45. Xiao, L.; Schultz, Z.D. Spectroscopic Imaging at the Nanoscale: Technologies and Recent Applications. *Anal. Chem.* **2018**, *90*, 440–458. [CrossRef] [PubMed]
46. Marcott, C.; Lo, M.; Kjoller, K.; Prater, C.; Gerrard, D.P. Applications of AFM-IR—Diverse Chemical Composition Analyses at Nanoscale Spatial Resolution. *Microsc. Today* **2012**, *20*, 16–21. [CrossRef]
47. Alonso-Lerma, B.; Barandiaran, L.; Ugarte, L.; Larraza, I.; Reifs, A.; Olmos-Juste, R.; Barruetabeña, N.; Amenabar, I.; Hillenbrand, R.; Eceiza, A.; et al. High performance crystalline nanocellulose using an ancestral endoglucanase. *Commun. Mater.* **2020**, *1*, 1–10. [CrossRef]
48. Hao, Y.; Cui, Z.; Yang, H.; Guo, G.; Liu, J.; Wang, Z.; Deniset-Besseau, A.; Remita, S. Adsorption of Cr (Vi) by Cellulose Adsorbent Prepared Using Ionic Liquid as a Green Homogeneous Reaction Medium. *Cellul. Chem. Technol.* **2018**, *52*, 485–494.
49. Igarashi, T.; Hoshi, M.; Nakamura, K.; Kaharu, T.; Murata, K.-I. Direct Observation of Bound Water on Cotton Surfaces by Atomic Force Microscopy and Atomic Force Microscopy–Infrared Spectroscopy. *J. Phys. Chem. C* **2020**, *124*, 4196–4201. [CrossRef]
50. Ugarte, L.; Santamaria-Echart, A.; Mastel, S.; Autore, M.; Hillenbrand, R.; Corcuera, M.A.; Eceiza, A. An alternative approach for the incorporation of cellulose nanocrystals in flexible polyurethane foams based on renewably sourced polyols. *Ind. Crop. Prod.* **2017**, *95*, 564–573. [CrossRef]
51. Nair, S.S.; Chen, H.; Peng, Y.; Huang, Y.; Yan, N. Polylactic Acid Biocomposites Reinforced with Nanocellulose Fibrils with High Lignin Content for Improved Mechanical, Thermal, and Barrier Properties. *ACS Sustain. Chem. Eng.* **2018**, *6*, 10058–10068. [CrossRef]
52. Jahng, J.; Yang, H.; Lee, E.S. Substructure imaging of heterogeneous nanomaterials with enhanced refractive index contrast by using a functionalized tip in photoinduced force microscopy. *Light Sci. Appl.* **2018**, *7*, 73. [CrossRef]
53. Ferreira, F.V.; Mariano, M.; Pinheiro, I.F.; Cazalini, E.M.; Souza, D.H.; Lepesqueur, L.S.; Koga-Ito, C.Y.; Gouveia, R.F.; Lona, L. Cellulose nanocrystal-based poly(butylene adipate-co-terephthalate) nanocomposites covered with antimicrobial silver thin films. *Polym. Eng. Sci.* **2019**, *59*, 356–365. [CrossRef]
54. Tibolla, H.; Pelissari, F.; Martins, J.; Lanzoni, E.; Vicente, A.; Menegalli, F.; Cunha, R. Banana starch nanocomposite with cellulose nanofibers isolated from banana peel by enzymatic treatment: In vitro cytotoxicity assessment. *Carbohydr. Polym.* **2019**, *207*, 169–179. [CrossRef] [PubMed]
55. Tibolla, H.; Czaikoski, A.; Pelissari, F.; Menegalli, F.; Cunha, R. Starch-based nanocomposites with cellulose nanofibers obtained from chemical and mechanical treatments. *Int. J. Biol. Macromol.* **2020**, *161*, 132–146. [CrossRef] [PubMed]

56. Marcott, C.; Lo, M.; Kjoller, K.; Prater, C.; Shetty, R.; Jakes, J.; Noda, I. Nanoscale Infrared Spectroscopy of Biopolymeric Materials. In *SAMPE Technical Conference Proceedings: Navigating the Global Landscape for the New Composites, Charleston, SC, USA, 22–25 October 2012 [CD-ROM]*; Society for the Advancement of Material and Process Engineering: Diamond Bar, CA, USA, 2012; 9p.

57. Soliman, M. Advanced Nanoscale Characterization of Plants and Plant-Derived Materials for Sustainable Agriculture and Renewable Energy. Ph.D. Thesis, University of Central Florida, Orlando, FL, USA, 2018.

58. Ciaffone, N. Interplay of Molecular and Nanoscale Behaviors in Biological Soft Matter. Master's Thesis, University of Central Florida, Orlando, FL, USA, 2018.

59. Piqueras, S.; Füchtner, S.; De Oliveira, R.R.; Gómez-Sánchez, A.; Jelavić, S.; Keplinger, T.; de Juan, A.; Thygesen, L.G. Understanding the Formation of Heartwood in Larch Using Synchrotron Infrared Imaging Combined with Multivariate Analysis and Atomic Force Microscope Infrared Spectroscopy. *Front. Plant Sci.* **2020**, *10*, 1701. [CrossRef] [PubMed]

60. Gusenbauer, C.; Jakob, D.S.; Xu, X.G.; Vezenov, D.V.; Cabane, É.; Konnerth, J. Nanoscale Chemical Features of the Natural Fibrous Material Wood. *Biomacromolecules* **2020**, *21*, 4244–4252. [CrossRef] [PubMed]

61. Patri, A.S.; Mostofian, B.; Pu, Y.; Ciaffone, N.; Soliman, M.; Smith, M.D.; Kumar, R.; Cheng, X.; Wyman, C.E.; Tetard, L.; et al. A Multifunctional Cosolvent Pair Reveals Molecular Principles of Biomass Deconstruction. *J. Am. Chem. Soc.* **2019**, *141*, 12545–12557. [CrossRef]

62. Wang, X.; Deng, Y.; Li, Y.; Kjoller, K.; Roy, A.; Wang, S. In situ identification of the molecular-scale interactions of phenol-formaldehyde resin and wood cell walls using infrared nanospectroscopy. *RSC Adv.* **2016**, *6*, 76318–76324. [CrossRef]

63. Wang, X.; Zhao, L.; Deng, Y.; Li, Y.; Wang, S. Effect of the penetration of isocyanates (pMDI) on the nanomechanics of wood cell wall evaluated by AFM-IR and nanoindentation (NI). *Holzforschung* **2018**, *72*, 301–309. [CrossRef]

64. Zancajo, V.M.R.; Lindtner, T.; Eisele, M.; Huber, A.J.; Elbaum, R.; Kneipp, J. FTIR Nanospectroscopy Shows Molecular Structures of Plant Biominerals and Cell Walls. *Anal. Chem.* **2020**, *92*, 13694–13701. [CrossRef] [PubMed]

65. Pereira, L.; Flores-Borges, D.N.; Bittencourt, P.R.; Mayer, J.L.; Kiyota, E.; Araújo, P.; Jansen, S.; Freitas, R.O.; Oliveira, R.S.; Mazzafera, P. Infrared Nanospectroscopy Reveals the Chemical Nature of Pit Membranes in Water-Conducting Cells of the Plant Xylem. *Plant Physiol.* **2018**, *177*, 1629–1638. [CrossRef]

66. Purohit, H.S.; Taylor, L.S. Miscibility of Itraconazole–Hydroxypropyl Methylcellulose Blends: Insights with High Resolution Analytical Methodologies. *Mol. Pharm.* **2015**, *12*, 4542–4553. [CrossRef]

67. Li, N.; Taylor, L.S. Nanoscale Infrared, Thermal, and Mechanical Characterization of Telaprevir–Polymer Miscibility in Amorphous Solid Dispersions Prepared by Solvent Evaporation. *Mol. Pharm.* **2016**, *13*, 1123–1136. [CrossRef] [PubMed]

68. Song, Y.-J.; Chen, C.-J.; Wu, Q.-L. Evolution of Functional Groups During the Preparation of Cellulose-Based Carbon Fibers Characterized by Nanoscale Infrared Spectroscopy. *New Carbon Mater.* **2019**, *34*, 296–301. (in Chinese). [CrossRef]

69. De Morais, V.B.; Corrêa, C.C.; Lanzoni, E.M.; Costa, C.A.R.; Bufon, C.C.B.; Santhiago, M. Wearable binary cooperative polypyrrole nanofilms for chemical mapping on skin. *J. Mater. Chem. A* **2019**, *7*, 5227–5233. [CrossRef]

70. Ruggeri, F.S.; Mannini, B.; Schmid, R.; Vendruscolo, M.; Knowles, T.P.J. Single molecule secondary structure determination of proteins through infrared absorption nanospectroscopy. *Nat. Commun.* **2020**, *11*, 1–9. [CrossRef]

71. Mester, L.; Govyadinov, A.A.; Chen, S.; Goikoetxea, M.; Hillenbrand, R. Subsurface chemical nanoidentification by nano-FTIR spectroscopy. *Nat. Commun.* **2020**, *11*, 3359. [CrossRef] [PubMed]

72. Fu, W.; Zhang, W. Hybrid AFM for Nanoscale Physicochemical Characterization: Recent Development and Emerging Applications. *Small* **2017**, *13*, 1603525. [CrossRef] [PubMed]

nanomaterials

Article

Flexible Nanocellulose/Lignosulfonates Ion-Conducting Separators for Polymer Electrolyte Fuel Cells

Carla Vilela [1],*, João D. Morais [1], Ana Cristina Q. Silva [1], Daniel Muñoz-Gil [2],†, Filipe M. L. Figueiredo [2], Armando J. D. Silvestre [1] and Carmen S. R. Freire [1],*

[1] Department of Chemistry, CICECO-Aveiro Institute of Materials, University of Aveiro, 3810-193 Aveiro, Portugal; jmorais@ua.pt (J.D.M.); ana.cristina.silva@ua.pt (A.C.Q.S.); armsil@ua.pt (A.J.D.S.)

[2] Department of Materials and Ceramic Engineering, CICECO-Aveiro Institute of Materials, University of Aveiro, 3810-193 Aveiro, Portugal; danielmg@ua.pt (D.M.-G.); lebre@ua.pt (F.M.L.F.)

* Correspondence: cvilela@ua.pt (C.V.); cfreire@ua.pt (C.S.R.F.)

† Present address: Institute of Ceramics and Glass, CSIC, Cantoblanco, 28049 Madrid, Spain.

Received: 31 July 2020; Accepted: 25 August 2020; Published: 29 August 2020

Abstract: The utilization of biobased materials for the fabrication of naturally derived ion-exchange membranes is breezing a path to sustainable separators for polymer electrolyte fuel cells (PEFCs). In this investigation, bacterial nanocellulose (BNC, a bacterial polysaccharide) and lignosulfonates (LS, a by-product of the sulfite pulping process), were blended by diffusion of an aqueous solution of the lignin derivative and of the natural-based cross-linker tannic acid into the wet BNC nanofibrous three-dimensional structure, to produce fully biobased ion-exchange membranes. These freestanding separators exhibited good thermal-oxidative stability of up to about 200 °C, in both inert and oxidative atmospheres (N_2 and O_2, respectively), high mechanical properties with a maximum Young's modulus of around 8.2 GPa, as well as good moisture-uptake capacity with a maximum value of ca. 78% after 48 h for the membrane with the higher LS content. Moreover, the combination of the conducting LS with the mechanically robust BNC conveyed ionic conductivity to the membranes, namely a maximum of 23 mS cm^{-1} at 94 °C and 98% relative humidity (RH) (in-plane configuration), that increased with increasing RH. Hence, these robust water-mediated ion conductors represent an environmentally friendly alternative to the conventional ion-exchange membranes for application in PEFCs.

Keywords: bacterial nanocellulose; lignosulfonates; mechanical performance; thermal-oxidative stability; ion-exchange membranes; biobased separators; ionic conductivity

---

1. Introduction

The increasing awareness toward clean energy and environmentally friendly materials is imposing a societal shift to meet the targets of the 2030 Agenda for Sustainable Development. Thus, the utilization of renewable raw materials for the development of the key components of fuel cells, which are efficient energy conversion technologies with zero-to-low emissions [1], is being explored to soften the impact of their production. Within the deluge of renewable raw materials, nanocellulose is one of the most interesting candidates to construct both the ion-exchange membrane and the electrodes for polymer electrolyte fuel cells (PEFCs), given its renewable nature, anisotropic shape, tailorable surface chemistry, and excellent mechanical properties, as recently reviewed by Vilela et al. [2]. In fact, bacterial nanocellulose (BNC), viz. the biotechnologically produced nanoscale form of cellulose [3,4], is particularly suitable, given its ability to be biosynthesized directly in the form of membranes with an adjustable size and shape, and also because of its unique mechanical performance. However, BNC presents a very low ionic conductivity [5,6], and therefore, the majority of the studies deal with either the chemical modification of BNC to introduce ionic moieties (e.g., sulfonic acid groups [7]) or the combination of BNC with synthetic polyelectrolytes (e.g., poly(bis[2-(methacryloyloxy)ethyl] phosphate) [8] and poly(4-styrene sulfonic acid) [9,10]) or ionomers (e.g., Nafion® [11,12]) that enable the transport of ions [2].

The demand for fully biobased ion-exchange membranes has already prompted the fabrication of, for example, ion-exchange membranes composed of chondroitin sulfate (a sulfated glycosaminoglycan) [13], cellulose nanocrystals obtained by acidic hydrolysis with sulfuric acid (a sulfated nanocellulose) [14], and fucoidan (a sulfated polysaccharide) combined with BNC [6]. In all these instances, the adsorption of water molecules assisted by the sulfate moieties produced paths for the structural diffusion of protons, which translated into separators with ionic conductivity. Lignosulfonates (LS), which are water-soluble anionic sulfonated lignin derivatives obtained as by-products of the sulfite pulping process [15,16], present a high content of sulfonate groups (sulfur content: 3.5–8.0 wt.% [17]) and, therefore, are strong contenders for non-expensive biobased ion conducting materials. Nevertheless, the high-water solubility of LS, and their non-film forming ability are major constraints for application in a PEFC that generates water and heat as reaction by-products. Although LS has already been blended with, for instance, poly(benzimidazole) [18], poly(sulfone) [19] and poly(styrene sulfonate)/nano-silica [20] for application in fuel cells, and BNC was previously biosynthesized in the presence of LS to assess the effect of the lignin derivative on the physical properties of BNC [21], the combination between LS and BNC has not yet been explored to fabricate biobased ion-exchange separators for PEFCs.

In this manner, the present study envisages the assembly and characterization of biobased separators composed of BNC and LS, for potential application as ion-exchange membranes. These naturally derived separators were assembled via diffusion of an aqueous solution of LS and tannic acid (TA, acting as a natural cross-linker), into the wet BNC nanofibrous three-dimensional structure. The resultant membrane separators were characterized in terms of structure (infrared spectroscopy), composition (energy dispersive X-ray spectrometry), morphology (scanning electron microscopy), thermal-oxidative stability (thermogravimetric analysis), mechanical performance (tensile tests), moisture-uptake capacity, ionic conductivity (impedance spectroscopy), and always compared with ion-exchange membranes reported in literature.

2. Materials and Methods

2.1. Chemicals and Materials

Lignosulfonic acid sodium salt (LS, Mw ~52,000 and Mn ~7000) and tannic acid (TA, $C_{76}H_{52}O_{46}$, from Chinese natural gall nuts) were acquired from Sigma-Aldrich (St. Louis, MO, USA). Ultrapure water (Type 1, 18.2 MΩ cm at 25 °C) was purified by a Simplicity® Water Purification System (Merck, Darmstadt, Germany). Additional chemicals or solvents were of laboratory grade.

Bacterial nanocellulose (BNC), entailing a three-dimensional network of nano- and micro-fibrils with 10–200 nm width, was biosynthesized in the form of wet membranes (99.5% of water) by the *Gluconacetobacter sacchari* bacterial strain [6].

2.2. Preparation of the BNC/LS-Based Membranes

BNC membranes in the wet state with a diameter of about 70 mm and 40% water content were placed on a Petri-dish having an aqueous solution of LS (2:1 and 4:3 mass fraction of BNC:LS, selected based on a previous study [6]) and TA (20% *w/w* relative to LS, chosen based on a previous study [6]), as summarized in Table 1. Following the complete absorption of the solutions (viz. 100% entrapment efficiency) at room temperature, the membranes were placed in a ventilated oven (Thermo Fisher

Scientific, Waltham, MA, USA) at 105 °C for 24 h to facilitate the thermal cross-linking during the drying process. All membrane separators were produced in triplicates and stored in desiccators.

Table 1. List of the prepared membranes with the corresponding composition and thickness values.

Membrane	$W_{BNC}:W_{LS}$ [a]	W_{LS}/V_{total} [mg cm^{-3}] [a]	Thickness [μm]
BNC	–	–	69 ± 11
BNC/LS_1	2:1	395 ± 16	75 ± 10
BNC/LS_2	4:3	547 ± 19	85 ± 9

[a] composition was estimated by considering the dry weight of BNC (W_{BNC}) and lignosulfonates (W_{LS}), and the volume of the membrane (V_{total}, determined by taking into account the diameter and thickness of the membranes); the values are expressed as mean ± standard deviation.

2.3. Characterization Methods

2.3.1. Thickness

A hand-held coolant proof digimatic micrometer MDC-25PX (Mitutoyo Corporation, Tokyo, Japan) was utilized to quantify the thickness at ten random sites of the membrane separators.

2.3.2. Attenuated Total Reflection-Fourier Transform Infrared (ATR-FTIR) Spectroscopy

A Perkin-Elmer FT-IR System Spectrum BX spectrophotometer (Perkin-Elmer Inc., Waltham, MA, USA) fitted out with a single horizontal Golden Gate ATR cell (Specac®, London, UK) was used to compute the ATR-FTIR spectra in the range of 600–4000 cm^{-1} at a resolution of 4 cm^{-1} over 32 scans.

2.3.3. Scanning Electron Microscopy (SEM) Combined with Energy Dispersive X-ray Spectroscopy (EDS)

An ultra-high-resolution field-emission HR-FESEM Hitachi SU-70 microscope (Hitachi High-Technologies Corporation, Tokyo, Japan), equipped with a microanalysis Bruker QUANTAX 400 detector for EDS (Bruker Nano GmbH, Berlin, Germany), was utilized to acquire micrographs of the membranes and evaluate their elemental chemical composition. Prior to analysis, the test specimens for surface and cross-section (fractured in liquid nitrogen) examination were put on a steel plate and coated with a carbon film.

2.3.4. Thermogravimetric Analysis (TGA)

A SETSYS Setaram TGA analyzer (SETARAM Instrumentation, Lyon, France) equipped with a platinum cell was used to assess the thermal stability. The test specimens were heated from 25 to 800 °C with a heating rate of 10 °C min^{-1} under two distinct atmospheres, namely nitrogen and oxygen.

2.3.5. Tensile Testing

A uniaxial Instron 5566 testing machine (Instron Corporation, Norwood, MA, USA) was utilized for the tensile tests in the traction mode at a crosshead velocity of 10 mm min^{-1} using a 500 N static load cell. The rectangular test specimens (50×10 mm^2) were formerly dried at 40 °C and all measurements were conducted on five replicates.

2.3.6. Moisture-Uptake Capacity

The moisture-uptake was quantified by putting the dry test specimens (20×20 mm^2) in a conditioned cabinet with 98% relative humidity (RH) (saturated potassium sulphate aqueous solution, $97.6 \pm 0.5\%$ [22]) at room temperature for 48 h. After taking the test specimens from the cabinet, the weight (W_w) was measured and the moisture-uptake capacity was determined as:

$$\text{Moisture} - \text{uptake} \ (\%) = (W_w - W_0) \times W_0^{-1} \times 100$$

where W_0 is the initial weight of the dry membrane.

2.3.7. Ionic Conductivity

Impedance spectroscopy (Agilent (County of Santa Clara, CA, USA) E4980A Precision LCR meter) was utilized to quantify the in-plane (IP) ionic conductivity (σ) at different temperature (40 °C to 94 °C) and RH (30% to 98%) conditions in an ACS Discovery DY110 climatic chamber (Angelantoni Test Technologies Srl, Massa Martana, Italy). The measurements were conducted on rectangular test specimens (ca. 15×5 mm^2) whereon two stripes of silver paste (Agar Scientific, Essex, UK) distancing ca. 10 mm were painted. Moreover, a pseudo 4-electrode configuration in a tubular sample holder was applied to guarantee full exposure of the test specimen surface to the controlled atmosphere and to give the necessary electrical contact between the test specimen and the LCR meter. The impedance spectra were collected between 20 Hz and 2×10^6 Hz with a test signal amplitude of 100 mV and analyzed with the ZView software (Version 2.6b, Scribner Associates (Southern Pines, NC, USA)) to calculate the Ohmic resistance (R) of the test specimen. The conductivity was then determined by the equation: $\sigma = L_0(R\delta\omega)^{-1}$, where L_0 is the distance between the two silver stripes, δ is the thickness of the membrane, and w is the width of the membrane.

3. Results and Discussion

3.1. Membrane Production and Characterization (Structure and Morphology)

Biobased ion-exchange separators were fabricated by combining a nanocellulose substrate, namely BNC, with a phenolic natural-based polyelectrolyte, namely LS, as outlined in Figure 1a. The straightforward diffusion of an aqueous solution of LS and TA (natural-based cross-linker) into the never-dried BNC nanofibrous three-dimensional structure, produced brownish BNC/LS-based membranes (Figure 1b), because of the signature dark brown color of the LS polyelectrolyte. BNC was picked for its in situ-moldability as a three-dimensional porous membrane, as well as good thermal stability and mechanical performance [23], while the lignin derivative (LS) was chosen for its high content of sulfonate groups ($-SO_3^-$) that facilitate ion motion and, hence, exhibits ionic conductivity [19]. On the other hand, the natural phenolic TA was selected for its cross-linking capability toward LS [24], as well as other neutral or charged macromolecules via physical or chemical interactions [25–27], to enable the retention of the water-soluble LS inside the wet BNC nanofibrous porous structure, as reported for other BNC-based membranes [6]. The resultant BNC/LS-based separators have two different compositions, namely 395 ± 16 mg of LS per cm^3 of membrane for the BNC/LS_1, and 547 ± 19 mg cm^{-3} for the BNC/LS_2, and thus, the thickness of the membranes increased with the increasing content of LS, as observed in Table 1.

The structural characterization of the membrane separators was carried out by ATR-FTIR vibrational spectroscopy. According to Figure 2a, the spectrum of the pristine bacterial polysaccharide displays the cellulose characteristic absorption bands at about 3341 cm^{-1} (O–H stretching), 2893 cm^{-1} (C–H stretching), 1314 cm^{-1} (O–H in plane bending), 1160 cm^{-1} (C–O–C antisymmetric stretching), and 1031 cm^{-1} (C–O stretching) [28,29]. The ATR-FTIR spectrum of LS (Figure 2a) exhibits the usual structural pattern of this lignin derivative with the presence of the absorption bands at around 3368 cm^{-1} (O–H stretching), 1570 cm^{-1} (C=C aromatic skeletal vibrations), 1410 cm^{-1} (C–O stretching), 1114, and 1040 cm^{-1} (S=O asymmetric and symmetric stretching), and 618 cm^{-1} (C–S stretching) [30,31]. The ATR-FTIR spectrum of TA (Figure 2a) presents the common absorption bands of an aromatic phenolic compound at about 3306 cm^{-1} (O–H stretching), 1700 cm^{-1} (C=O stretching), 1606 cm^{-1} (C–C aromatic stretching), 1308 cm^{-1} (C$_{ar}$–OC stretching, C$_{ar}$–O–H in-plane bending, C–C aromatic stretching), and 1174 cm^{-1} (O–CO and C$_{ar}$–CO stretching, C$_{ar}$–O–H in-plane bending) [32].

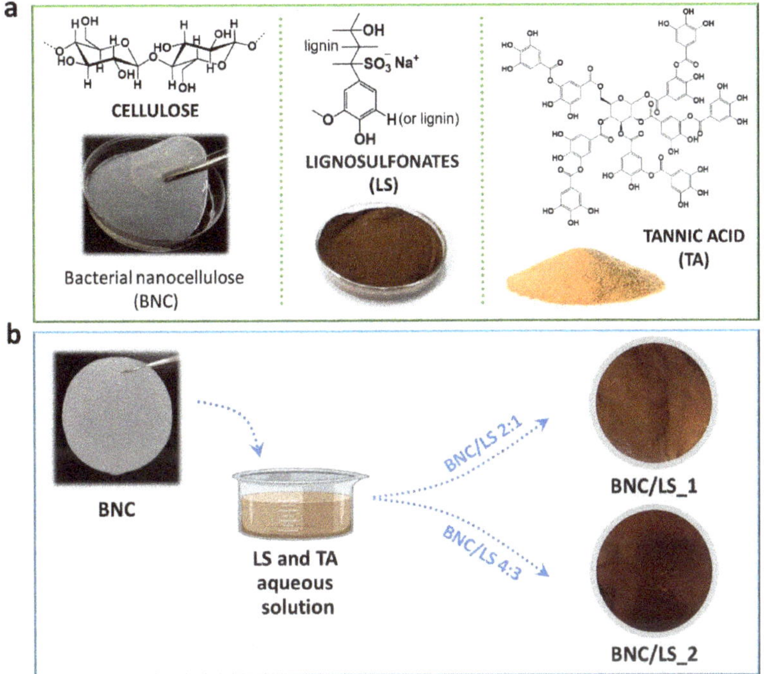

Figure 1. (**a**) Chemical structure, and photographs of the precursors, namely bacterial nanocellulose (BNC, wet membrane), lignosulfonates (LS, powder) and tannic acid (TA, powder), and (**b**) route for the production of the BNC/LS-based membranes and the photographs of the respective dry membranes.

Additionally, the FTIR-ATR spectra of both BNC/LS-based membranes (Figure 2a) show similarities with those of their precursors, particularly with BNC and LS. Although most of the absorption bands of LS and TA are entirely overlapped with the vibrations of the dominant component, viz. BNC, the efficient inclusion of LS into the BNC nanofibrous three-dimensional structure was clearly confirmed.

The elemental chemical composition of the BNC/LS-based membranes was assessed by EDS analysis, as depicted in Figure 2b. The EDS spectra of the two biobased separators confirm the existence of BNC and LS through the detection of the sulfur (S), sodium (Na), oxygen (O), and carbon (C) peaks at 2.31, 1.04, 0.51, and 0.27 keV, respectively. Predictably, the presence of sodium shows the bound of the sulfonate moieties to the Na cations. The analysis of both EDS spectra demonstrates that the sulfur content of the BNC/LS_1 membrane is lower than that of BNC/LS_2, which is in line with the relative contents of BNC and LS used in their preparation. Furthermore, the EDS mapping of sulfur at the surface of both membrane separators showed an appreciable content and a uniform distribution of the element (Figure 2b, inset images with sulfur element in red color). Therefore, the lignin derivative was effectively incorporated into the BNC porous three-dimensional network, as already verified by infrared spectroscopy.

The morphology of the membrane separators was examined by SEM with the surface and cross-section micrographs of BNC and BNC/LS-based membranes shown in Figure 2c. A glimpse over the micrographs of the surface and cross-section of the pristine BNC membrane evidences the morphological traits of this nanocellulose substrate, namely the well-known nanofibrillar and lamellar microstructure [23]. In Figure 2c, it is further noticeable that the inclusion of the sulfonated lignin derivative camouflaged the nanofibrils and occupied the lamellar spaces of the BNC porous network, which is particularly notorious for the membrane containing the higher LS content (547 ± 19 mg cm^{-3},

BNC/LS_2, Table 1). This behavior is quite common as in fact documented for other partially and fully biobased BNC-based ion-exchange separators [5,6,8].

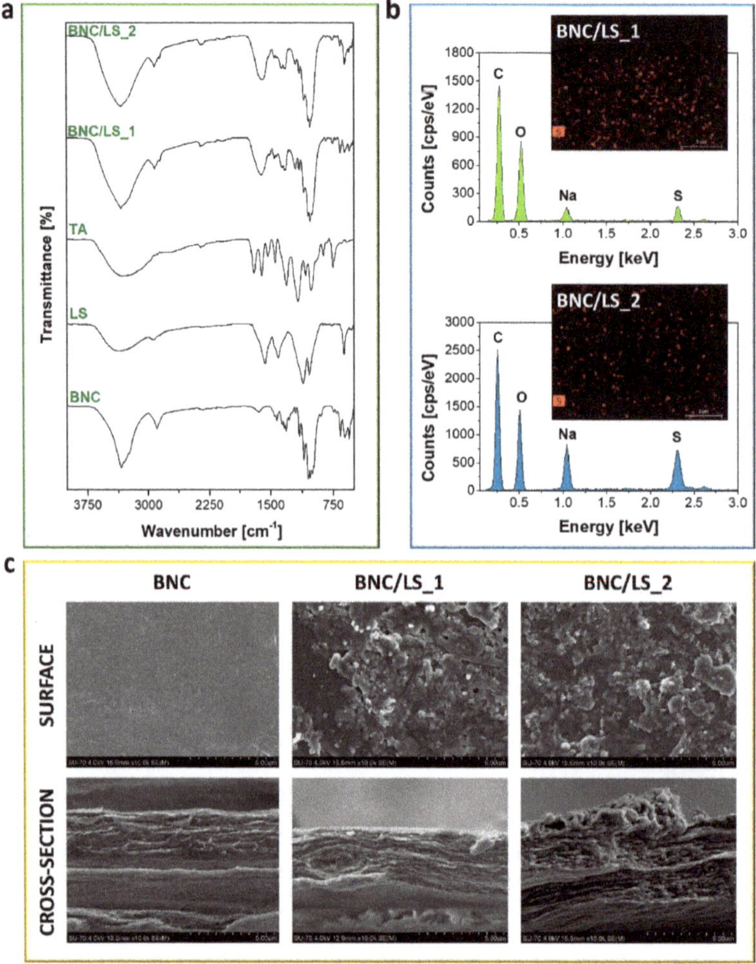

Figure 2. (**a**) Attenuated total reflection-Fourier transform infrared (ATR-FTIR) spectra of BNC, LS, TA, and BNC/LS-based membranes, (**b**) EDS spectra and mapping (scale bar: 3 μm) of the cross-section of the two BNC/LS membranes, and (**c**) SEM micrographs of the surface and cross-section of the pristine BNC and BNC/LS-based membranes (×10.0 k magnification).

3.2. Thermal and Mechanical Properties

Thermogravimetric analysis (TGA) was utilized to investigate the thermal-oxidative stability of the BNC/LS-based membranes, along with the pristine BNC and LS samples. The degradation profile of BNC (Figure 3a) under inert atmosphere follows a single weight-loss stage with maximum decomposition temperature of 350 °C, as a consequence of the pyrolysis of the cellulose skeleton [33]. On the other hand, the LS exhibits a degradation profile composed of two weight-loss stages (Figure 3a), while the initial water evaporation below 100 °C. The first weight-loss step occurs at a maximum decomposition temperature of 270 °C, allocated to the pyrolysis of oxygen-containing groups, while the second stage appears at a maximum decomposition temperature of 695 °C and is ascribed to the loss of the remaining O_2-containing groups on carbon edges [34].

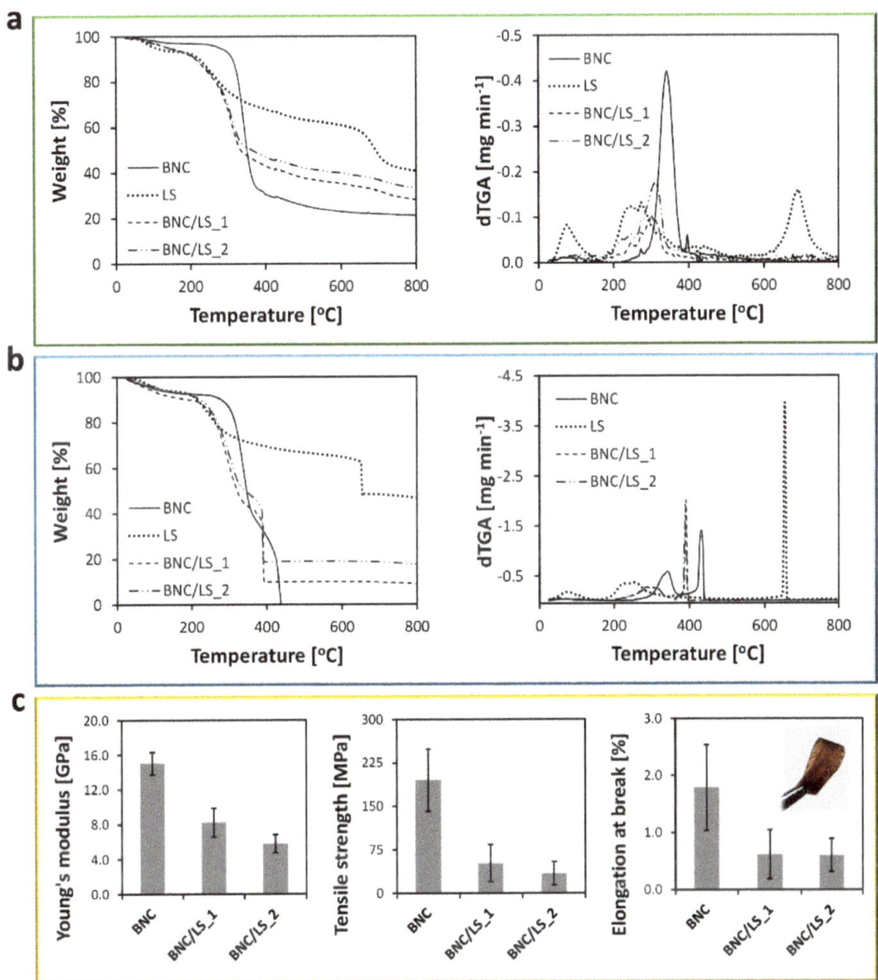

Figure 3. (**a**,**b**) Thermograms (left) with the respective derivatives (right) of pristine BNC, LS, and BNC/LS-based membranes under (**a**) N_2 (inert) and (**b**) O_2 (oxidative) atmospheres, and (**c**) tensile tests data: Young's modulus (left), tensile strength (middle), and elongation at break (right, the inset photograph corresponds to BNC/LS_1) of the pristine BNC and BNC/LS-based membranes.

The thermograms of the two BNC/LS-based membranes under N_2 atmosphere (Figure 3a) present a similar profile with one weight-loss step with maximum rate of decomposition temperatures at 314 °C for BNC/LS_1 and 318 °C for BNC/LS_2, and a residue at 800 °C that increased with the LS content from 28% for BNC/LS_1 to 32% for BNC/LS_2. So, the combination between LS and BNC created membrane separators with lower thermal stability when compared with the pristine BNC, because of the presence of the less thermally stable, and amorphous LS polyelectrolyte (Figure 3a). Similar results were obtained for membranes composed of BNC and fucoidan, where the inclusion of the sulphated polysaccharide into the BNC network also yielded materials with lower thermal stability than the pristine BNC [6].

When applied in a PEFC, the BNC/LS-based membranes will have to withstand an oxidative environment, therefore, their thermal-oxidative stability was also measured under oxidative atmosphere (Figure 3b). For the pristine BNC membrane, the process in oxygen is marked by a two-stage degradation profile with maximum rate of decomposition temperatures at ca. 340 and 433 °C, reaching a complete degradation with no residue [35,36]. For the LS powder, the thermogram is composed by two weight-loss stages with maximum rate of decomposition temperatures at around 246 and 653 °C, while the loss of water below 100 °C (loss of ca. 5%), leaving a residue of about 46% at 800 °C. This profile is roughly equivalent to the process under N_2 atmosphere, both in terms of temperatures and the total residue content at 800 °C.

The TGA tracing of the BNC/LS-based membranes (Figure 3b), contrary to the process in N_2, displays two weight-loss stages with maximum rate of decomposition temperatures at around 310 and 391 °C for BNC/LS_1, and 318 and 389 °C for BNC/LS_2, with residues of 9 and 18% of the initial mass at 800 °C for BNC/LS_1 and BNC/LS_2, respectively. Although the BNC/LS-based membranes present a lower thermal stability than the commercial Nafion® ionomer used in PEFCs (ca. 290 °C [12]), both membranes are thermally stable at least up to 200 °C in both inert and oxidative atmospheres. Hence, their thermal-oxidative profile does not jeopardize the envisioned application as ion-exchange membranes for PEFCs that operate under temperatures below 100 °C.

The mechanical performance of the BNC/LS-based membranes was investigated by tensile tests and the respective data are compiled in Figure 3c. The membrane composed solely of BNC (thickness: 69 ± 11 μm) presents values of Young's modulus of 15.0 ± 1.3 GPa, tensile strength of 195 ± 54 MPa and elongation at break of 1.8 ± 0.8%, which are in tune with data reported elsewhere [36]. No values were obtained for the LS since this amorphous polymer is not a film-forming material. On the other hand, the incorporation of LS into the nanostructured BNC, produced membrane separators with lower mechanical performance when compared with the pristine BNC, but with the advantage that the pure LS does not form free-standing films. The BNC/LS_1 membrane exhibits values of Young's modulus of 8.2 ± 1.6 GPa, tensile strength of 52 ± 32 MPa, and elongation at break of 0.6 ± 0.4%, whereas the BNC/LS_2 membrane presents a Young's modulus of 5.8 ± 1.1 GPa, tensile strength of 34 ± 20 MPa and elongation at break of 0.6 ± 0.3% (Figure 3c). Although the reduction of these three mechanical parameters translates into less stiffer materials, the membranes are bendable (see the inset photograph in Figure 3c) and still adequate for application as ion separators for PEFCs.

When compared with the ion-exchange membranes reported in literature, the BNC/LS-based membrane separators developed in the present study possess lower mechanical performance than, e.g., the membranes constituted by BNC combined with poly(4-styrene sulfonic acid), which is a synthetic polyelectrolyte containing sulfonic acid moieties [10]. Nevertheless, their mechanical performance is comparable, for example, with those of membranes composed of poly(benzimidazole) and LS [18], and poly(styrene sulfonate), LS and nano-silica [20], but most importantly they present superior mechanical properties than the reference benchmark Nafion® membrane with a Young's modulus of 0.25 GPa and tensile strength of 43 MPa [37].

3.3. Moisture-Uptake Capacity and Ionic Conductivity

The moisture-uptake capacity of the BNC/LS-based membranes, as well as of the pristine BNC membrane and LS powder, was estimated by positioning the materials in a chamber with controlled humidity, namely 98% relative humidity (RH), during 48 h. Predictably, both BNC and LS have the aptitude to absorb environmental humidity, although with distinct values, i.e., 21.8 ± 2.1% for BNC [6] and 214.0 ± 5.7% for LS, given their different natures. The two BNC/LS-based membranes absorbed moisture with values of 55.6 ± 2.4% for BNC/LS_1 and 78.0 ± 3.7% for BNC/LS_2, as epitomized in Figure 4a. Furthermore, both membranes guarded their mechanical integrity after the moisture absorption tests, confirming the good wet-dimensional stability of BNC. The values obtained for both separators are superior than the water absorption of the benchmark Nafion® (~38%, when fully hydrated at 100 °C for 1 h) [37]. So, the inclusion of LS is clearly beneficial toward the augment of the moisture-uptake capacity that actively affects the ionic conductivity [38]. Understandably, the data show that the adsorption of water molecules is supported by the sulfonate groups, which will produce, without a doubt, paths for the structural diffusion of ions [2].

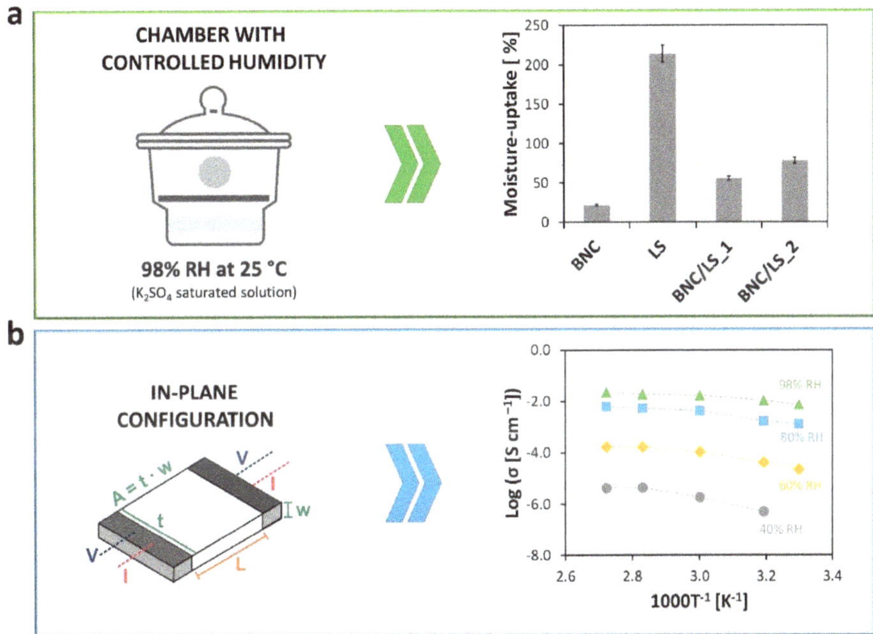

Figure 4. (a) Moisture-uptake capacity of the pristine BNC, LS, and BNC/LS-based membranes after 48 h at room temperature and 98% relative humidity (RH) in a chamber with controlled humidity, and (b) Arrhenius-type plot of the ionic conductivity (σ) of BNC/LS_2 membrane at different RH (40%, 60%, 80%, and 98%) in the in-plane configuration [5].

In this perspective, the ionic conductivity (σ) of the Na-form membrane with the higher LS content (i.e., BNC/LS_2) was measured by impedance spectroscopy in the in-plane configuration (Figure 4b). Previous studies have shown that the pristine BNC is a poor ionic conductor (63 μS cm^{-1} at 98% RH and 94 °C) [5,6], just like the other nanofibrillar form of cellulose, viz., cellulose nanofibrils (50 μS cm^{-1} at nominal 100% RH and 100 °C) [14]. Nevertheless, when combined with a lignin derivative with a polyelectrolytic nature, namely LS, the BNC membranes turns into an ion-conducting material. In fact, the BNC/LS_2 membrane exhibits ionic conductivity that increases with the rise in relative humidity (40–98% RH) and temperature (30–94 °C), as depicted in the Arrhenius-type plot of Figure 4b. Inevitably, the RH is the factor that primarily affects the ionic conductivity in the case of water-mediated ionic conductors [2].

According to Table 2, the ionic conductivity increases up to five orders of magnitude when the RH rises from 40% to 98%, and only one order of magnitude when the temperature varies from 30 to 94 °C. For instance, the conductivity increased from 5.0×10^{-7} S cm^{-1} at 40% RH and 40 °C to 1.1×10^{-2} S cm^{-1} at 98% RH and 40 °C, while it only increased from 2.2×10^{-5} S cm^{-1} at 30 °C and 60% RH to 1.7×10^{-4} S cm^{-1} at 40 °C and 60% RH. Furthermore, the maximum ionic conductivity was reached at elevated humidity and temperature conditions (98% RH and 94 °C) with a value of 23 mS cm^{-1}.

After the measurements at 98% RH and 94 °C, no apparent degradation is visible, and the membrane withheld its mechanical integrity.

Table 2. Ionic conductivity values obtained for the BNC/LS_2 membrane at different temperature (30–94 °C) and relative humidity (RH, 40–98%).

Temperature [°C]	Ionic Conductivity [S cm^{-1}]			
	40% RH	60% RH	80% RH	98% RH
30	–	2.2×10^{-5}	1.3×10^{-3}	7.3×10^{-3}
40	5.0×10^{-7}	4.1×10^{-5}	1.7×10^{-3}	1.1×10^{-2}
60	1.8×10^{-6}	1.1×10^{-4}	4.2×10^{-3}	1.7×10^{-2}
80	4.3×10^{-6}	1.7×10^{-4}	5.4×10^{-3}	1.9×10^{-2}
94	4.1×10^{-6}	1.7×10^{-4}	6.1×10^{-3}	2.3×10^{-2}

Herein, it should be noted that for lower RH the ionic conductivity is focused essentially on Na$^+$ ion transport, while at higher RH there is probably some exchange of Na$^+$ ions by protons, and the ionic conductivity might be a binary cation system focused on proton and Na$^+$ ion conduction. In this context, Saito et al. [39] studied the mechanisms of ion and water transport of the commercial Nafion® membranes in the H- and Na-forms. According to the authors, both forms presented ionic conductivity that improved with increasing water content of the membranes, and the ionic conductivity of the H-form membrane (~150 mS cm^{-1}, fully hydrate state, 25 °C) is far greater than that of the Na-form membrane (~35 mS cm^{-1}, fully hydrate state, 25 °C).

In addition, the dependency of the ionic conductivity on temperature shows a slight bending that indicates a reduction of the apparent activation energy with the increase in temperature (Figure 4b). Such type of Vogel–Tammann–Fulcher (VTF) behavior indicates a role of the segmental motion in assisting ion transport particularly under low humidity when sodium transport is expected to be relatively more important. At high humidity (80% RH and more), the proton concentration is the highest and the conductivity is expected to be essentially due to proton transport in the aqueous domains formed in the structure. In these conditions, the dependency of the ionic conductivity on temperature can be described by the Arrhenius relationship: $\sigma = \sigma_0 e^{-E_a(RT)^{-1}}$ (σ_0—pre-exponential term, E_a—activation energy, R—gas constant, T—absolute temperature [9,40]), with an estimated E_a for ion transport of ca. 26 kJ mol^{-1} at 80% RH and 18 kJ mol^{-1} at 98% RH. These values paralleled with those typically recorded for BNC [6] and Nafion® [41,42], suggesting they are governed by similar ion conduction mechanism(s) [43,44].

Albeit the lower ionic conductivity of BNC/LS_2 when compared with the standard Nafion® [42,45], the BNC/LS_2 membrane developed in the present study displays conductivity values that are comparable and, in some cases, even higher than other partially and fully biobased ion-exchange membranes reported in literature, as outlined in Table 3. Those examples include the membranes composed of CNCs, chitosan, and poly(vinyl alcohol) with a highest conductivity of 0.642 mS cm^{-1} (25 °C, fully hydrated) [46], the BNC/fucoidan membrane with a highest conductivity of 1.6 mS cm^{-1} (94 °C, 98% RH) [6], and the pure cellulose nanocrystals (CNCs) membrane with a highest conductivity of 2.5 mS cm^{-1} (90 °C, nominal 100% RH) [14]. On the other hand, it should be pointed out that there is a partially biobased membrane with an ionic conductivity that can reach up to 406 mS cm^{-1} (25 °C, fully hydrated), but only for the reason that three materials with high ionic conductivity, namely lignosulfonates, poly(styrene sulfonate), and nano-silica, are combined [20]. Nevertheless, these membranes composed of BNC and LS, with moderate ionic conductivity under variable temperature and humidity conditions, good mechanical performance, thermal-oxidative stability under inert and oxidative environments, and dimensional stability under humid conditions, show potential as an eco-friendly alternative of ion conductors for application in PEFCs.

Table 3. Examples of partially and fully biobased ion-exchange membranes reported in literature and compared with the present study.

	Components [a]	Conductivity [a]	Ref.
Partially biobased			
	CNCs/CH/PVA	0.642 mS cm^{-1} (IP, 25 °C, fully hydrated)	[46]
	CNFs/PBI	66.6 mS cm^{-1} (IP, 140 °C)	[47]
	BNC/Nafion®	140 mS cm^{-1} (IP, 94 °C, 98% RH)	[12]
	BNC/PSSA	185 mS cm^{-1} (IP, 94 °C, 98% RH)	[5]
	BNC/PMOEP	100 mS cm^{-1} (TP, 80 °C, 98% RH)	[35]
	BNC/P(bisMEP)	30 mS cm^{-1} (IP, 80 °C, 98% RH)	[8]
	k-carrageenan/IL	186 mS cm^{-1} (IP, 60 °C, 98% RH)	[48]
	LS/PBI	187 mS cm^{-1} (IP, 160 °C, anhydrous)	[18]
	LS/PSSA/nano-silica	406 mS cm^{-1} (TP, 25 °C, fully hydrated)	[20]
Fully biobased			
	BNC/LS	23 mS cm^{-1} (IP, 94 °C, 98% RH)	Present study
	BNC/Fucoidan	1.6 mS cm^{-1} (TP, 94 °C, 98% RH)	[6]
	CNCs	2.5 mS cm^{-1} (TP, 90 °C, 100% RH) 4.6 mS cm^{-1} (TP, 120 °C, 100% RH)	[14]
	Chondroitin sulfate/citric acid	37 mS cm^{-1} (TP, 25 °C, 98% RH)	[13]

[a] BNC: bacterial nanocellulose, CH: chitosan, CNCs: cellulose nanocrystals, IL: ionic liquid (1-butyl-3-methyl-1*H*-imidazolium chloride ([Bmim]Cl)), IP: in-plane, LS: lignosulfonate, P(bisMEP): poly(bis[2-(methacryloyloxy)ethyl] phosphate), PBI: poly(benzimidazole), PMOEP: poly(methacryloyloxyethyl phosphate), PSSA: poly(4-styrene sulfonic acid), PVA: poly(vinyl alcohol), RH: relative humidity, TP: through-plane.

4. Conclusions

The present study underlines the combination of two naturally derived polymeric materials, namely nanocellulose and a lignin derivative (lignosulfonates), to build fully biobased and ease-to-prepare ion-exchange membrane separators for utilization in polymer electrolyte fuel cells. The obtained freestanding membranes manifested appropriate thermal-oxidative stability up to about 200 °C in either N_2 (inert) or O_2 (oxidative) atmospheres, elevated mechanical properties with a maximum Young's modulus of 8.2 GPa, along with good moisture-uptake capacity with a maximum value of ca. 78% after 48 h. Additionally, the blend of the conducting lignosulfonates with the mechanically robust bacterial nanocellulose granted a maximum ionic conductivity of 23 mS cm^{-1} at 94 °C and 98% RH to the membrane with the highest LS content. Therefore, these BNC/LS water-mediated ion conductors, with good mechanical performance, thermal-oxidative stability, and water-uptake capacity and whose conductivity is actively linked to humidification, can be employed as eco-friendly substitutes to ion-exchange membranes for application in PEFCs.

Author Contributions: Conceptualization, C.V. and C.S.R.F.; investigation, C.V., J.D.M., A.C.Q.S., and D.M.-G.; writing—original draft preparation, C.V.; writing—review and editing, C.V., J.D.M., A.C.Q.S., D.M.-G., F.M.L.F., A.J.D.S., and C.S.R.F.; supervision, C.V. and C.S.R.F.; funding acquisition, F.M.L.F., A.J.D.S., and C.S.R.F. All authors have read and agreed to the published version of the manuscript.

Funding: This work was developed within the scope of the project CICECO-Aveiro Institute of Materials, UIDB/50011/2020 & UIDP/50011/2020, financed by national funds through the Portuguese Foundation for Science and Technology (FCT)/MCTES. FCT is also acknowledged for the doctoral grant to A.C.Q.S. (SFRH/BD/140230/2018) and the research contract under Scientific Employment Stimulus to C.V. (CEECIND/00263/2018). Additional funding to D.M.G. and F.M.L.F. through project UniRCell (SAICTPAC/0032/2015, POCI-01-0145-FEDER-016422) co-financed by FCT/MEC and FEDER under the PT2020 Partnership Agreement.

Conflicts of Interest: The authors declare no conflict of interest.

References

1. Scofield, M.E.; Liu, H.; Wong, S.S. A concise guide to sustainable PEMFCs: Recent advances in improving both oxygen reduction catalysts and proton exchange membranes. *Chem. Soc. Rev.* **2015**, *44*, 5836–5860. [CrossRef] [PubMed]

2. Vilela, C.; Silvestre, A.J.D.; Figueiredo, F.M.L.; Freire, C.S.R. Nanocellulose-based materials as components of polymer electrolyte fuel cells. *J. Mater. Chem. A* **2019**, *7*, 20045–20074. [CrossRef]

3. Wang, J.; Tavakoli, J.; Tang, Y. Bacterial cellulose production, properties and applications with different culture methods—A review. *Carbohydr. Polym.* **2019**, *219*, 63–76. [CrossRef] [PubMed]

4. Jacek, P.; Dourado, F.; Gama, M.; Bielecki, S. Molecular aspects of bacterial nanocellulose biosynthesis. *Microb. Biotechnol.* **2019**, *12*, 633–649. [CrossRef] [PubMed]

5. Gadim, T.D.; Loureiro, F.J.; Vilela, C.; Rosero-Navarro, N.C.; Silvestre, A.; Freire, C.; Figueiredo, F.M. Protonic conductivity and fuel cell tests of nanocomposite membranes based on bacterial cellulose. *Electrochim. Acta* **2017**, *233*, 52–61. [CrossRef]

6. Vilela, C.; Silva, A.C.; Domingues, E.M.; Gonçalves, G.; Martins, M.A.; Figueiredo, F.M.L.; Santos, S.A.O.; Freire, C. Conductive polysaccharides-based proton-exchange membranes for fuel cell applications: The case of bacterial cellulose and fucoidan. *Carbohydr. Polym.* **2019**, *230*, 115604. [CrossRef]

7. Yue, L.; Xie, Y.; Zheng, Y.; He, W.; Guo, S.; Sun, Y.; Zhang, T.; Liu, S. Sulfonated bacterial cellulose/polyaniline composite membrane for use as gel polymer electrolyte. *Compos. Sci. Technol.* **2017**, *145*, 122–131. [CrossRef]

8. Vilela, C.; Martins, A.P.C.; Sousa, N.; Silvestre, A.; Figueiredo, F.M.L.; Freire, C. Poly(bis[2-(methacryloyloxy) ethyl] phosphate)/Bacterial Cellulose Nanocomposites: Preparation, Characterization and Application as Polymer Electrolyte Membranes. *Appl. Sci.* **2018**, *8*, 1145. [CrossRef]

9. Gadim, T.D.O.; Figueiredo, A.G.P.R.; Rosero-Navarro, N.C.; Vilela, C.; Gamelas, J.A.F.; Barros-Timmons, A.; Neto, C.P.; Silvestre, A.; Freire, C.; Figueiredo, F.M.L. Nanostructured Bacterial Cellulose–Poly(4-styrene sulfonic acid) Composite Membranes with High Storage Modulus and Protonic Conductivity. *ACS Appl. Mater. Interfaces* **2014**, *6*, 7864–7875. [CrossRef]

10. Vilela, C.; Cordeiro, D.M.; Boas, J.V.; Barbosa, P.; Nolasco, M.; Vaz, P.D.; Rudić, S.; Ribeiro-Claro, P.; Silvestre, A.J.; Oliveira, V.B.; et al. Poly(4-styrene sulfonic acid)/bacterial cellulose membranes: Electrochemical performance in a single-chamber microbial fuel cell. *Bioresour. Technol. Rep.* **2020**, *9*, 100376. [CrossRef]

11. Jiang, G.-P.; Zhang, J.; Qiao, J.; Jiang, Y.-M.; Zarrin, H.; Chen, Z.; Hong, F.N. Bacterial nanocellulose/Nafion composite membranes for low temperature polymer electrolyte fuel cells. *J. Power Sources* **2015**, *273*, 697–706. [CrossRef]

12. Gadim, T.D.; Vilela, C.; Loureiro, F.J.; Silvestre, A.; Freire, C.; Figueiredo, F.M. Nafion® and nanocellulose: A partnership for greener polymer electrolyte membranes. *Ind. Crop. Prod.* **2016**, *93*, 212–218. [CrossRef]

13. Santos, F.M.; Barbosa, P.C.; Pereira, R.F.; Silva, M.M.; Gonçalves, H.M.; Nunes, S.C.; Figueiredo, F.L.; Valente, A.J.M.; Bermudez, V.D.Z. Proton conducting electrolytes composed of chondroitin sulfate polysaccharide and citric acid. *Eur. Polym. J.* **2020**, *124*, 109453. [CrossRef]

14. Bayer, T.; Cunning, B.V.; Selyanchyn, R.; Nishihara, M.; Fujikawa, S.; Sasaki, K.; Lyth, S.M. High Temperature Proton Conduction in Nanocellulose Membranes: Paper Fuel Cells. *Chem. Mater.* **2016**, *28*, 4805–4814. [CrossRef]

15. Holladay, J.E.; White, J.F.; Bozell, J.J.; Johnson, D. *Top Value-Added Chemicals from Biomass Volume II—Results of Screening for Potential Candidates from Biorefinery Lignin*; U.S. Department of Energy: Washington, DC, USA, 2007.

16. Moreno, A.; Sipponen, M.H. Lignin-based smart materials: A roadmap to processing and synthesis for current and future applications. *Mater. Horiz.* **2020**. [CrossRef]

17. Aro, T.; Fatehi, P. Production and Application of Lignosulfonates and Sulfonated Lignin. *ChemSusChem* **2017**, *10*, 1861–1877. [CrossRef]

18. Barati, S.; Abdollahi, M.; Ghazi, M.M.; Khoshandam, B. High temperature proton exchange porous membranes based on polybenzimidazole/lignosulfonate blends: Preparation, morphology and physical and proton conductivity properties. *Int. J. Hydrogen Energy* **2019**, *44*, 30440–30453. [CrossRef]

19. Zhang, X.; Benavente, J.; Garcia-Valls, R. Lignin-based membranes for electrolyte transference. *J. Power Sources* **2005**, *145*, 292–297. [CrossRef]

20. Gonggo, S.T.; Bundjali, B.; Hariyawati, K.; Arcana, I.M. The influence of nano-silica on properties of sulfonated polystyrene-lignosulfonate membranes as proton exchange membranes for direct methanol fuel cell application. *Adv. Polym. Technol.* **2017**, *37*, 1859–1867. [CrossRef]

21. Keshk, S.M.A.S. Physical properties of bacterial cellulose sheets produced in presence of lignosulfonate. *Enzym. Microb. Technol.* **2006**, *40*, 9–12. [CrossRef]

22. Greenspan, L. Humidity fixed points of binary saturated aqueous solutions. *J. Res. Natl. Bur. Stand. Sect. Phys. Chem.* **1977**, *81*, 89–96. [CrossRef]

23. Klemm, D.; Cranston, E.D.; Fischer, D.; Gama, M.; Kedzior, S.A.; Kralisch, D.; Kramer, F.; Kondo, T.; Lindström, T.; Nietzsche, S.; et al. Nanocellulose as a natural source for groundbreaking applications in materials science: Today's state. *Mater. Today* **2018**, *21*, 720–748. [CrossRef]

24. Ghazali, N.A.; Naganawa, S.; Masuda, Y. Feasibility Study of Tannin-Lignosulfonate Drilling Fluid System for Drilling Geothermal Prospect. In Proceedings of the 43rd Workshop on Geothermal Reservoir Engineering, Standford University, Stanford, CA, USA, 12–14 February 2018. SGP-TR-213.

25. Chen, W.; Li, N.; Ma, Y.; Minus, M.L.; Benson, K.; Lu, X.; Wang, X.; Ling, X.; Zhu, H. Superstrong and Tough Hydrogel through Physical Cross-Linking and Molecular Alignment. *Biomacromolecules* **2019**, *20*, 4476–4484. [CrossRef] [PubMed]

26. Erel-Unal, I.; Sukhishvili, S.A. Hydrogen-Bonded Multilayers of a Neutral Polymer and a Polyphenol. *Macromolecules* **2008**, *41*, 3962–3970. [CrossRef]

27. Fan, H.; Wang, L.; Feng, X.; Bu, Y.; Wu, D.; Jin, Z. Supramolecular Hydrogel Formation Based on Tannic Acid. *Macromolecules* **2017**, *50*, 666–676. [CrossRef]

28. Foster, E.J.; Moon, R.J.; Agarwal, U.P.; Bortner, M.J.; Bras, J.; Camarero-Espinosa, S.; Chan, K.J.; Clift, M.J.D.; Cranston, E.D.; Eichhorn, S.J.; et al. Current characterization methods for cellulose nanomaterials. *Chem. Soc. Rev.* **2018**, *47*, 2609–2679. [CrossRef]

29. Vilela, C.; Oliveira, H.; Almeida, A.; Silvestre, A.; Freire, C. Nanocellulose-based antifungal nanocomposites against the polymorphic fungus Candida albicans. *Carbohydr. Polym.* **2019**, *217*, 207–216. [CrossRef]

30. Shen, Q.; Zhang, T.; Zhu, M.-F. A comparison of the surface properties of lignin and sulfonated lignins by FTIR spectroscopy and wicking technique. *Colloids Surf. Physicochem. Eng. Asp.* **2008**, *320*, 57–60. [CrossRef]

31. Bellamy, L.J. *The Infrared Spectra of Complex Molecules*, 3rd ed.; Chapman and Hall Ltd.: London, UK, 1975; ISBN 0412138506.

32. Ranoszek-Soliwoda, K.; Tomaszewska, E.; Socha, E.; Krzyczmonik, P.; Ignaczak, A.; Orlowski, P.; Krzyzowska, M.; Celichowski, G.; Grobelny, J. The role of tannic acid and sodium citrate in the synthesis of silver nanoparticles. *J. Nanoparticle Res.* **2017**, *19*, 273. [CrossRef]

33. Pa'e, N.; Salehudin, M.H.; Hassan, N.D.; Marsin, A.M.; Muhamad, I.I. Thermal behavior of bacterial cellulose-based hydrogels with other composites and related instrumental analysis. In *Cellulose-Based Superabsorbent Hydrogels*; Polymers and Polymeric Composites: A Reference Series; Mondal, M.I.H., Ibrahim, M., Eds.; Springer: Berlin/Heidelberg, Germany, 2019; pp. 763–787.

34. Yao, J.; Odelius, K.; Hakkarainen, M. Carbonized lignosulfonate-based porous nanocomposites for adsorption of environmental contaminants. *Funct. Compos. Mater.* **2020**, *1*, 1–12.

35. Vilela, C.; Gadim, T.D.O.; Silvestre, A.; Freire, C.; Figueiredo, F.M.L. Nanocellulose/poly(methacryloyloxyethyl phosphate) composites as proton separator materials. *Cellulose* **2016**, *23*, 3677–3689. [CrossRef]

36. Vilela, C.; Sousa, N.; Pinto, R.J.B.; Silvestre, A.; Figueiredo, F.M.; Freire, C. Exploiting poly(ionic liquids) and nanocellulose for the development of bio-based anion-exchange membranes. *Biomass Bioenergy* **2017**, *100*, 116–125. [CrossRef]

37. DuPontTM Nafion®N115, N117, N1110—Ion Exchange Materials, Product Bulletin P-12. Available online: http://fuelcellearth.com/pdf/nafion-N115-N117-N1110.pdf (accessed on 7 July 2020).

38. Bose, S.; Kuila, T.; Nguyen, T.X.H.; Kim, N.H.; Lau, K.-T.; Lee, J.H. Polymer membranes for high temperature proton exchange membrane fuel cell: Recent advances and challenges. *Prog. Polym. Sci.* **2011**, *36*, 813–843. [CrossRef]

39. Saito, M.; Arimura, N.; Hayamizu, K.; Okada, T. Mechanisms of Ion and Water Transport in Perfluorosulfonated Ionomer Membranes for Fuel Cells. *J. Phys. Chem. B* **2004**, *108*, 16064–16070. [CrossRef]

40. Petrowsky, M.; Frech, R. Temperature Dependence of Ion Transport: The Compensated Arrhenius Equation. *J. Phys. Chem. B* **2009**, *113*, 5996–6000. [CrossRef]

41. Rosero-Navarro, N.C.; Domingues, E.M.; Sousa, N.; Ferreira, P.; Figueiredo, F.M.L. Protonic conductivity and viscoelastic behaviour of Nafion®membranes with periodic mesoporous organosilica fillers. *Int. J. Hydrog. Energy* **2014**, *39*, 5338–5349. [CrossRef]

42. Rosero-Navarro, N.C.; Domingues, E.M.; Sousa, N.; Ferreira, P.; Figueiredo, F.M. Meso-structured organosilicas as fillers for Nafion®membranes. *Solid State Ion.* **2014**, *262*, 324–327. [CrossRef]

43. Eikerling, M.; Kornyshev, A.A.; Kuznetsov, A.M.; Ulstrup, J.; Walbran, S. Mechanisms of Proton Conductance in Polymer Electrolyte Membranes. *J. Phys. Chem. B* **2001**, *105*, 3646–3662. [CrossRef]

44. Miyake, T.; Rolandi, M. Grotthuss mechanisms: From proton transport in proton wires to bioprotonic devices. *J. Phys. Condens. Matter* **2015**, *28*, 023001. [CrossRef]

45. Cele, N.; Ray, S.S. Recent Progress on Nafion-Based Nanocomposite Membranes for Fuel Cell Applications. *Macromol. Mater. Eng.* **2009**, *294*, 719–738. [CrossRef]

46. Gaur, S.S.; Dhar, P.; Sonowal, A.; Sharma, A.; Kumar, A.; Katiyar, V. Thermo-mechanically stable sustainable polymer based solid electrolyte membranes for direct methanol fuel cell applications. *J. Membr. Sci.* **2017**, *526*, 348–354. [CrossRef]

47. Esmaielzadeh, S.; Ahmadizadegan, H. Construction of proton exchange membranes under ultrasonic irradiation based on novel fluorine functionalizing sulfonated polybenzimidazole/cellulose/silica bionanocomposite. *Ultrason. Sonochem.* **2018**, *41*, 641–650. [CrossRef] [PubMed]

48. Nunes, S.C.; Pereira, R.F.P.; Sousa, N.; Silva, M.M.; Almeida, P.C.; Figueiredo, F.M.L.; Bermudez, V.D.Z. Eco-Friendly Red Seaweed-Derived Electrolytes for Electrochemical Devices. *Adv. Sustain. Syst.* **2017**, *1*, 1700070. [CrossRef]

 *nanomaterials*

Article

Influence of Drying Method and Argon Plasma Modification of Bacterial Nanocellulose on Keratinocyte Adhesion and Growth

Anna Kutová [1,*,†], Lubica Staňková [2,*,†], Kristýna Vejvodová [1], Ondřej Kvítek [1], Barbora Vokatá [3], Dominik Fajstavr [1], Zdeňka Kolská [4], Antonín Brož [2], Lucie Bačáková [2] and Václav Švorčík [1]

[1] Department of Solid State Engineering, Institute of Chemical Technology, Prague, Technická 5, 16628 Prague, Czech Republic; vejvodok@vscht.cz (K.V.); kviteko@vscht.cz (O.K.); fajstavd@vscht.cz (D.F.); svorcikv@vscht.cz (V.Š.)

[2] Department of Biomaterials and Tissue Engineering, Institute of Physiology of the Czech Academy of Sciences, Vídeňská 1038, 14220 Prague, Czech Republic; antonin.broz@fgu.cas.cz (A.B.); lucie.bacakova@fgu.cas.cz (L.B.)

[3] Department of Biochemistry and Microbiology, Institute of Chemical Technology, Prague, Technická 5, 16628 Prague, Czech Republic; vokataa@vscht.cz

[4] Materials Centre of Usti nad Labem, Faculty of Science, J. E. Purkyně University, Pasteurova 15, 40096 Ústí nad Labem, Czech Republic; zdenka.kolska@ujep.cz

* Correspondence: kutovaa@vscht.cz (A.K.); lubica.stankova@fgu.cas.cz (L.S.); Tel.: +420-220-445-142 (A.K.)

† Equal contributions.

Citation: Kutová, A.; Staňková, L.; Vejvodová, K.; Kvítek, O.; Vokatá, B.; Fajstavr, D.; Kolská, Z.; Brož, A.; Bačáková, L.; Švorčík, V. Influence of Drying Method and Argon Plasma Modification of Bacterial Nanocellulose on Keratinocyte Adhesion and Growth. *Nanomaterials* **2021**, *11*, 1916. https://doi.org/10.3390/nano11081916

Academic Editor: Takuya Kitaoka

Received: 29 June 2021
Accepted: 23 July 2021
Published: 26 July 2021

Publisher's Note: MDPI stays neutral with regard to jurisdictional claims in published maps and institutional affiliations.

Abstract: Due to its nanostructure, bacterial nanocellulose (BC) has several advantages over plant cellulose, but it exhibits weak cell adhesion. To overcome this drawback, we studied the drying method of BC and subsequent argon plasma modification (PM). BC hydrogels were prepared using the *Komagataeibacter sucrofermentans* (ATCC 700178) bacteria strain. The hydrogels were transformed into solid samples via air-drying (BC-AD) or lyophilization (BC-L). The sample surfaces were then modified by argon plasma. SEM revealed that compared to BC-AD, the BC-L samples maintained their nanostructure and had higher porosity. After PM, the contact angle decreased while the porosity increased. XPS showed that the O/C ratio was higher after PM. The cell culture experiments revealed that the initial adhesion of human keratinocytes (HaCaT) was supported better on BC-L, while the subsequent growth of these cells and final cell population density were higher on BC-AD. The PM improved the final colonization of both BC-L and BC-AD with HaCaT, leading to formation of continuous cell layers. Our work indicates that the surface modification of BC renders this material highly promising for skin tissue engineering and wound healing.

Keywords: bacterial nanocellulose; lyophilization; plasma modification; cell adhesion

1. Introduction

Bacterial (or microbial) nanocellulose (BC) has been known for more than two thousand years as a by-product of the kombucha tea fermentation process [1], although the first person to scientifically observe and describe this material was A. J. Brown in 1886. During his work with *Bacterium acetum*, he described it as a translucent jelly-like material that occurs on the surface of the cultivation fluid and proved it to be cellulose [2]. Nowadays, several gram-negative aerobic rod-like bacteria genera with high acid tolerance producing BC are known, especially the most efficient producer of BC—*Komagataeibacter* (formally known as *Gluconacetobacter* or *Acetobacter*) [3–5]. On the liquid surface, the bacteria form BC that protects them from dry-out, irradiation, lack of oxygen, and pathogens [6]. Many examples of scientific articles describing the production of BC using various bacterial species and subspecies, media composition (carbon and nitrogen source, pH), and reaction conditions leading to materials of various shapes and properties can be found in literature [7–9].

Although BC has the same chemical composition as plant cellulose (PC), it differs significantly in its other properties. BC is obtained in higher purity since there is no need for the removal of other plant polymers [10]. It consists of fibres that are thinner than 100 nm (compared to PC having fibres the size of around 30 μm [11]) so it can be classified as nanocellulose. This nanostructure provides BC with its remarkable properties such as high porosity, specific surface area, crystallinity (60–80% [12,13]), water-holding capacity, permeability for gases and liquids, and excellent mechanical properties [14,15]. The degree of polymerization of BC is usually between 2000 and 6000 [16].

Bacterial nanocellulose and its composites have already been used in many fields of industry such as fashion [17], papermaking and packaging [18,19], audio membranes [20], air purification membranes [21], and cosmetics [22]. BC has also been used in medicinal applications [23] such as drug delivery vehicles (e.g., propranolol [24], ibuprofen, lidocaine [25], doxorubicin [26]), ophthalmology [27,28], and regenerative medicine [29–32]. Nowadays, there are several BC-based commercially available wound dressings [33] that are supposed to treat ulcers, burns, or chronic wounds such as Bionext® [34], XCell® [35], Nexfill®, Nanoderm™ [36] and Dermafill® [37]. These materials induce epithelialization with no need for everyday re-dressing [23] thanks to their permeability for air and liquids [12,38]. BC has gained popularity in these medicinal applications thanks to its non-pyrogenicity, non-toxicity, biocompatibility, similarity to soft skin tissue, and ability to provide an optimal three-dimensional substrate for cell attachment [12,23,39].

Although BC exhibits good biocompatibility [40] leading to the above mentioned medicinal applications, it exhibits quite poor cell adhesion, which can be improved by several different methods: (1) immobilization of various adhesion proteins, (2) preparation of BC-based composites with various biomolecules such as gelatin or collagen, (3) by plasma surface modification, (4) or by tailoring the surface properties such as porosity or morphology [41].

Plasma modification (PM) is a method that can be used to modify the biomaterial surface into the depth of about 1 nm while maintaining the properties of the bulk material. This leads to a biomaterial with altered surface properties such as morphology, chemical composition, and hydrophilicity but with preserved mechanical properties and functionality [42,43]. Alteration of these surface properties can lead to enhanced biocompatibility and cell adhesion, since the modified surfaces provide a better cell support, e.g., by improved absorption of cell adhesion-mediating proteins [10]. BC has been so far modified with nitrogen [44], oxygen and fluoromethane [45] plasmas. The cited studies observed that plasma-modified BC contains higher concentration of functional groups improving the cell adhesion, has higher porosity and lower water contact angle. These changes resulted in improved cell adhesion.

For PM and manipulation of the surface properties generally, it is important that the sample is properly dried. After harvesting and washing the BC hydrogel, the excessive water can be removed by several drying methods: air drying, oven drying, draining with water-absorbing material, supercritical drying, or lyophilization. Generally, lyophilization and supercritical drying are milder drying methods; the material maintains its nanostructure, and therefore shows a higher water swelling ratio and porosity [6,8,9,46,47].

In this work the influence of different drying methods of BC with subsequent PM on the surface properties, morphology, and keratinocyte adhesion was studied (for the diagram of the work see Figure 1). Since the cell adhesion on native BC is quite poor, we chose the Ar⁺ PM as a cheap, quick, proven, and easy method to control the surface properties. The plasma-modified samples exhibited a higher porosity due to the etching and a lower contact angle. This method has been shown to be able to enhance the cell adhesion on BC.

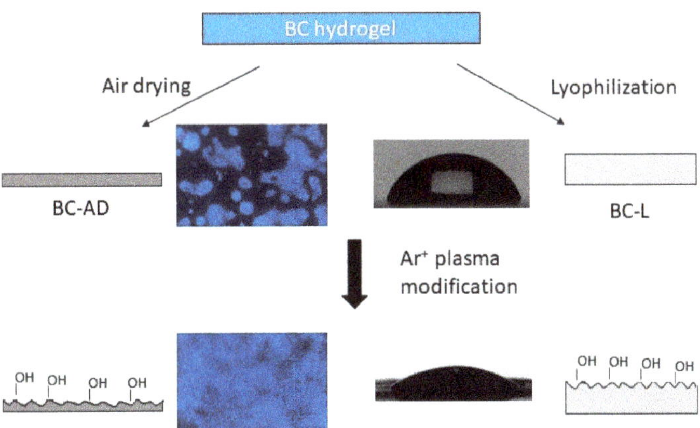

Figure 1. Diagram of the mechanism for improving the cell adhesion of bacterial nanocellulose.

2. Materials and Methods

2.1. Preparation of BC Foils

Bacterial nanocellulose was produced by *Komagataeibacter sucrofermentans* (Leibniz-Institut DSMZ, Braunschweig, Germany, DSM 15973). Cultivation was carried out in Hestrin-Shramm [48] culture medium, consisting of D-glucose (20 g/L), disodium hydrogen phosphate dodecahydrate (6.8 g/L), special peptone (5 g/L), yeast extract (5 g/L), and citric acid monohydrate (1.3 g/L), pH 5.8. Cultivation lasted for at least 7 days at 28 °C in Erlenmeyer flask statically. Purification of the nanocellulose from the bacteria and medium residue was performed by rinsing in boiling 0.1 M NaOH two times and then rinsing in boiling distilled water two times. To fully remove the bacteria residue, we washed the samples in 10% (m/m) solution of SDS and trypsin solution. Washed BC hydrogels were solidified via air-drying on PTFE foil (AD) or lyophilization (L) (FreeZone 2.5, Labconco, Kansas City, MO, USA) for at least 24 h. Circular samples with a diameter of 16 mm were then cut from the dried BC foils.

2.2. Plasma Modification of BC Foils

The surface of the solid samples was modified in a direct (glow, diode) Ar⁺ plasma discharge on Balzers SCD 050 device (BAL-TEC, Balzers, Lichtenstein). The conditions were set as follows: gas purity 99.997%, pressure of 7 Pa, electrode distance of 55 mm, electrode area of 48 cm², chamber volume of approx. 1 dm³, plasma volume of 0.24 dm³, electrical current of 15 mA, and voltage of 680 V. The samples were modified from both sides for several different exposure times (60 s, 240 s, and 480 s). The samples were then named BC-AD 60 s, BC-AD 240 s, and BC-AD 480 s and BC-L 60 s, BC-L 240 s, and BC-L 480 s for BC-AD and BC-L, respectively.

2.3. Methods of Analysis

The thickness of the dried BC foils was measured with digital caliper micrometer QuantuMike IP65 (0–25 mm, 0.001 mm, Mitutoyo, Kawasaki, Japan). The measurement was repeated for at least 20 different areas on each sample.

Chemical composition of materials and evaluation of chemical changes after freeze-drying and plasma modification were measured by Fourier transform infrared spectroscopy (FTIR—ThermoFisher, Nicolet iS5 with iD7 attenuated total reflection accessory with diamond crystal, Waltham, MA, USA). The spectra were obtained as an average from 128 measurement cycles with a spectral range of 600–4000 cm⁻¹, and 1 cm⁻¹ data interval. The changes of surface composition were also studied by X-ray photoelectron spectroscopy, ESCAProbe P (Omicron Nanotechnology, GmbH, Taunusstein, Germany) device was used for these measurements using a monochromatic energy source at 1486.7 eV. The exposed and analyzed area was 2×3 mm^2 and the spectra were obtained with 0.05 eV energy step. CasaXPS software was used to analyze the spectra.

The surface morphology of BC nanofibers was studied using a dual-beam focused ion beam-scanning electron microscope with a FEG electron gun (FIB-SEM TESCAN LYRA3GMU, Brno, Czech Republic). The sample morphology was investigated using SEM at an acceleration voltage of 7 keV and a deceleration voltage of 5 keV. Prior to the measurement, the samples were sputtered with a thin film of platinum and attached to the sample holder with silver paint to avoid surface charging during the measurements.

Water contact angles of BC samples were measured by the sessile drop method in order to evaluate the changes in surface hydrophilicity by fully automated goniometer DSA100 (KRÜSS GmbH, Hamburg, Germany). A 2 µL droplet of distilled water was produced on a capillary tip above the studied surface of BC and after 5 s it was carried to the surface. The whole sequence was recorded on video at a recording speed of 50 fps. The contact angle was calculated from the captured frame where the sessile drop just spread over the surface to its highest diameter to eliminate the influence of the water being absorbed into the material.

Surface area and pore volume were determined by N_2 adsorption/desorption isotherms. Samples were degassed at room temperature for 24 h. After that, adsorption and desorption isotherms were measured with nitrogen (N_2, Linde, 99.999% purity) using Quantachrome Instruments NOVA3200 (Anton Paar GmbH, Graz, Austria). All samples were measured three times with an experimental error of less than 5%. Five-point Brunauer-Emmett-Teller (BET) analysis has been applied to determine the total surface area and a 40-points Barrett-Joyner-Halenda (BJH) model was used for determining the pore volume.

Gravimetric analysis was used to get the material loss after plasma modification. Samples were weighed before and after plasma modification at least five times each on the UMX2 ultra-microbalance system (Mettler Toledo, Greifenses, Switzerland). The weight loss Δm was calculated using this equation: $\Delta m = (m_0 - m_1)/m_0$, where m_0 and m_1 stand for weight before and after modification, respectively.

2.4. Cell Model and Culture Conditions

The human keratinocytes of the line HaCaT, purchased from CLS Cell Lines Service (Eppelheim, Germany), were cultivated in Dulbecco's Modified Eagle's Medium (DMEM; Sigma-Aldrich Co., St Louis, MO, USA) with 10% of fetal bovine serum (FBS; Sebak GmbH, Aidenbach, Germany) and 40 µg/mL of gentamicin (Novartis International AG, Basel, Switzerland).

Selected circular BC samples (BC-AD and BC-L, either unmodified or modified with plasma for 240 s) were sterilized with in an autoclave (121 °C, 23 min, 101.3 kPa) and inserted into the wells of 24-well cell culture polystyrene plates (TPP, Trasadingen, Switzerland). The cells were seeded on the samples at a density of approximately 15,000 cells/cm^2 (i.e., 30,000 cells/well) into 1.5 mL of the cell culture medium (mentioned above) per well. The cells were then cultivated for three time periods (1, 4, and 7 days) at 37 °C and in a humidified air atmosphere with 5% CO_2. Tissue culture polystyrene (PS) wells were used as reference material.

2.5. Evaluation of the Cell Number, Morphology, and Viability

The number and morphology of HaCaT cells on BC samples and control PS wells were evaluated on days 1, 3, and 7 after seeding. First, the cells were rinsed with phosphate-buffered saline (PBS) and were fixed with -20 °C cold ethanol for 5 min. Then, the cells were incubated with a combination of fluorescent dyes diluted in PBS, namely Hoechst 33258, which stains the cell nuclei (5 µg/mL; Sigma-Aldrich, Schnelldorf, Germany), and Texas Red C_2-maleimide, which stains the proteins of the cell membrane and cytoplasm (20 ng/mL; Thermo Fisher Scientific, Waltham, MA, USA), for 1 h at room temperature in the dark. Images of the cells were taken using an epifluorescence microscope (IX 51; Olympus, Tokyo, Japan; objective 4×), equipped with a digital camera (DP 70). On day 1 after seeding, the number of human HaCaT keratinocytes was evaluated by direct counting on the images taken under the fluorescence microscope using the ImageJ software. In the following days (days 4 and 7), when the direct cell counting was disabled by the increasing cell density and cell overlapping, the cell number was estimated indirectly by measuring the intensity of fluorescence of Hoechst 33258-stained cells on microphotographs, taken at the same exposure time for all experimental groups, using ImageJ software.

The viability of HaCaT cells was measured on day 4 after seeding by a trypan blue-exclusion test in an automated Vi-Cell XR Cell Viability Analyser (Beckman Coulter, Indianapolis, IN, USA) from four parallel samples of each experimental group. Before the analysis, the cells were detached from the material by incubation in a trypsin-EDTA solution (0.05% trypsin, 0.02% EDTA, Sigma-Adrich, Schnelldorf, Germany) for 8 min at 37 °C in a humidified air atmosphere with 5% CO_2.

2.6. Statistics

Quantitative data are presented as arithmetic mean ± standard deviation values (S.D.) or standard error of the mean (S.E.M.) from three or more independent samples for each experimental group. Statistical significance was evaluated using SigmaPlot 14.0, analysis of variance, Student–Newman–Keuls method, or nonparametric Kruskal–Wallis test. Values of $p \leq 0.05$ were considered significant.

3. Results and Discussion

3.1. Production of BC Pellicles

Bacteria of the *Komagataeibacter sucrofermentans* (ATCC 700178, DSM 15973) strain were cultivated for at least 7 days to obtain BC pellicles. Literature sources suggest 7-day cultivation to be sufficient for BC production due to the declining carbon source after this time [49–52]. After harvesting and washing as reported above, we obtained transparent hydrogel of total mass approx. 23 mg out of 100 mL of medium. The yield did not increase further with time. This hydrogel had an uneven surface (Figure 2a). We suggest this to be caused by the non-homogeneous distribution of bacteria cells in the pellicle during cultivation, which leads to thicker (cloudy) and thinner (transparent) parts. These heterogeneities are also visible on the lyophilized BC (BC-L) (Figure 2c), where the cloudy parts are changed into white opaque material, while the transparent parts remain as they were. The air drying (AD) was carried out on a hydrophobic PTFE foil to prevent the material from sticking to the drying pad (Figure 2b). This BC-AD did not show any visible heterogeneities in the surface. These two drying methods also showed differences between the thickness of the material, which was (10.55 ± 3.99) µm for BC-AD, (19.45 ± 6.41) µm for the transparent part of BC-L, and (51.45 ± 13.53) µm for the white parts of BC-L. These results follow the trend suggested by Illa [47] and Vasconcellos [46] that the freeze-drying preserves the morphology of the original hydrogel and therefore the film thickness is significantly higher compared to AD (resp. oven dried in their case). Zeng [8] did not observe such significant differences between AD and L. This could be caused by selection of different bacterial strain which resulted in different morphology of the material. Compared to Illa [47], we did not observe differences between the fiber diameters. Our material showed fiber diameter of (58.22 ± 14.47) nm for BC-AD, (65.21 ± 15.00) nm for the transparent part of BC-L, and (54.88 ± 14.57) nm for the white part of BC-L. These results are in accordance with those observed by Zeng [8]. For the following experiments circular samples with diameter of 16 mm were cut. There was roughly the same amount of white and transparent parts in these BC-L circles.

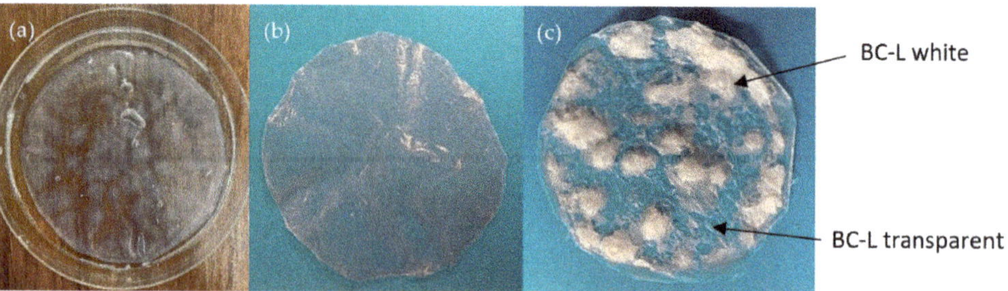

Figure 2. (**a**) Harvested and washed bacterial nanocellulose (BC) pellicle; (**b**) air-dried BC pellicle; (**c**) lyophilized BC pellicle.

3.2. Surface Modification of BC

In this work we modified both BC-AD and BC-L samples with Ar$^+$ plasma with the power of 7 W. We used three different exposure times to evaluate the effects of the modification process. Samples were modified from both sides for easier further manipulation. The differences were evaluated by gravimetric analysis, FTIR, and XPS to examine the composition changes, SEM for morphology changes, and contact angle to determine the hydrophilicity.

The Ar$^+$ PM leads to surface ablation of the samples (plasma etching). The rate of the ablation depends on the chemical composition of the polymer chain and properties of the used plasma discharge (plasma type, composition, discharge power, and exposure time) [53]. Almost all the modified samples showed degradation to various extents. Sample edges were burned and turned brown-blackish, but the central parts of the samples remained intact. This was, however, not the case of the samples modified for 480 s that showed signs of degradation over the whole area of the samples. The middle part degraded with much higher intensity, burned holes in the samples were observed (Figure 3). Gravimetric analysis (Table 1) showed that these samples have significantly higher weight loss. Therefore the samples with 480 plasma exposure time were not further examined, since the degradation leads to very inhomogeneous sample surface. Overall, the BC-AD samples showed higher rate of weight loss. This could be caused by the removal of larger parts of the material during PM due to their more compact and at the same time brittle nature.

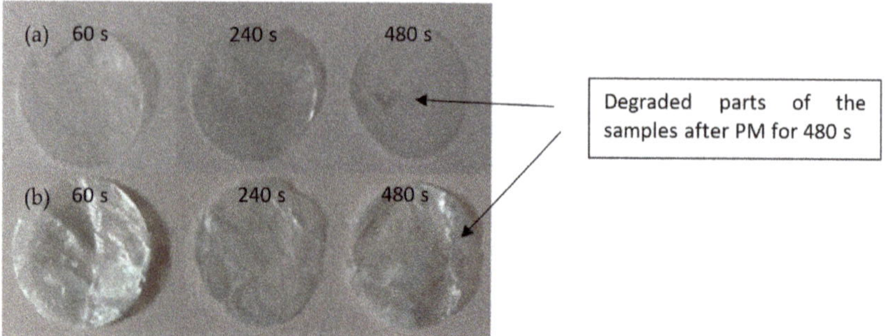

Figure 3. Samples after plasmatic modification for various exposure times (60–480 s): (**a**) BC-AD, (**b**) BC-L. The samples with 480 s PM shows signs of degradation.

Table 1. Gravimetric analysis of the BC weight loss after PM for different exposure times.

Sample	Weight Loss [%]
BC-AD 60 s	3.20 ± 0.36
BC-AD 240 s	14.26 ± 0.61
BC-AD 480 s	27.89 ± 2.88
BC-L 60 s	2.40 ± 0.45
BC-L 240 s	6.80 ± 1.27
BC-L 480 s	12.64 ± 3.00

3.3. Chemical Composition of BC

FTIR spectra of the BC-AD and BC-L samples unmodified and after 60 s and 240 s PM are shown in Figure 4. Spectral interpretation and band assignment in cellulose is considered to be somewhat problematic due to the dominant role of inter- and intra-chain hydrogen bond interactions that lead to numerous combination vibrations [54]. However, some general features of the material can be ascribed to the absorption bands in the present spectra. The spectra of the prepared BC correspond to cellulose I structure with both Iα and Iβ components being present, which is confirmed by the presence of weak absorption bands at both 744 and 710 cm^{-1}. This can be assigned to the glucose ring deformations, compounded with glycosidic bond bending of the respective cellulose structure modifications [54].

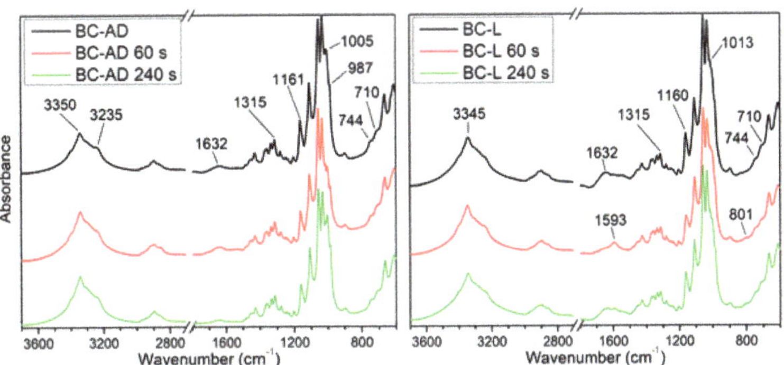

Figure 4. FTIR absorbance spectra of the BC-AD and BC-L unmodified and plasma-modified samples.

The OH band around 3350 cm^{-1} combines the vibrations of the hydrogen bonds in cellulose. The absorption at 3235 cm^{-1}, which is slightly stronger in the case of the BC-AD sample is usually attributed to the 2O-H···6O-H···3O hydrogen bond group [55].

The absorption in the 1700–1500 cm^{-1} range relates to adsorbed water [56]. While all the BC-AD samples show a single weak absorption at 1632 cm^{-1}, in the BC-L samples the absorption is noticeably stronger and the absorbed range is wider, in certain cases with a dominant second absorption maximum at 1593 cm^{-1}. This indicates there is a higher amount of adsorbed water in the BC-L samples and the water binds to the cellulose structure differently than in the air-dried samples.

The absorption band at 1315 cm^{-1}, which can be assigned to C-O-H bending vibrations [57], is relatively stronger in the BC-AD samples. The only absorption band where a consistent shift with the PM can be observed was the 1160 cm^{-1} which shifts about 2 cm^{-1} to lower wavenumber in samples after PM. Moreover, this band appeared on average about 20% weaker in those samples. These changes are in the scale of the whole spectrum rather insignificant; however, this can be expected, since PM mainly influences the very surface layer of the material and the FTIR signal is obtained over the sample depth of several micrometers. Nonetheless, the variation of this absorption band could mean the glycosidic bond and the cellulose chains have been to some degree disrupted by the PM. The air-dried samples also show weaker maxima at 1005 and 987 cm^{-1} (bands related to vibrations of C-O bonds in the hydroxyl groups), while in the lyophilized samples these absorptions show merely as shoulders of the stronger bands. Especially in the case of the air-dried samples, the absorption at 1005 cm^{-1} shows very high variability, even within repeated measurements on a single sample. In the case of the BC-L samples with PM, a weak band was observed around 800 cm^{-1}. Absorptions in this region are usually assigned to deformations of the glucose ring coupled with bending of the glycosidic bond. Therefore, this could indicate the influence of PM on the glycosidic bond in the BC samples.

XPS measurements showed the changes in the surface composition, especially in the carbon and oxygen content. The samples were somewhat contaminated with Al, Si, F, S, Na, and N from the plasma vacuum chamber, and the bacteria and medium residues. The results shown in Table 2 are recalculated without the contaminating elements. The expected value for O/C from the chemical structure of cellulose is 0.83. For unmodified samples, this value is slightly lower, which is caused by the residues of media and bacteria on the surface of the material. The most significant result is the large increase in oxygen content after PM of BC-L 240 s. This could be caused by the disruption of the cellulose structure leading to the formation of highly reactive species on the surface, which react with ambient oxygen after the exposure of the modified surface to atmosphere. The increase in oxygen content after PM for BC-AD is also apparent.

Table 2. Carbon and oxygen atomic concentrations of BC samples measured by XPS with 0° detection angle.

Sample	C (%)	O (%)	O/C
BC-AD	55.80	44.20	0.6124
BC-AD 60 s	58.39	41.61	0.7126
BC-AD 240 s	59.49	40.51	0.6809
BC-L	57.24	42.76	0.6595
BC-L 60 s	59.36	40.64	0.6847
BC-L 240 s	49.69	50.31	1.0125

3.4. Surface Morphology of BC

The SEM images in Figure 5 show the nanostructured morphology of unmodified BC-AD and BC-L (transparent and white part) samples. The differences between BC-AD and BC-L samples can be seen in the porosity of the material. For the L samples, the fibers are more spread out. This corresponds with the fact that the lyophilized pellicles are thicker. Illa [47], who compared lyophilization and oven drying, suggested that this phenomenon is caused by the free hydroxyl groups that can form secondary bonds. The mobility of the amorphous regions during freezing is reduced and therefore the morphology is preserved. However, during oven drying the thermal energy maintains the mobility of the amorphous chains and therefore the morphology collapses, the fibers come closer together, and those samples are thinner. Air drying at room temperature could have similar effect on the material morphology. Together with the smaller thickness of the material compared to BC-L, SEM shows that BC-AD has a more rugged surface structure with the bent fibers turned perpendicular to its surface. The differences between SEM images of BC-L transparent and BC-L white were studied as well. Those nanofibers were found to be indistinguishable in the SEM images for both samples. The plasma-modified samples did not show any differences in these areas either, so we can suppose that the thickness of the lyophilized pellicles does not affect the nanofibrillar structure and it is determined by the bacterial origin of the material.

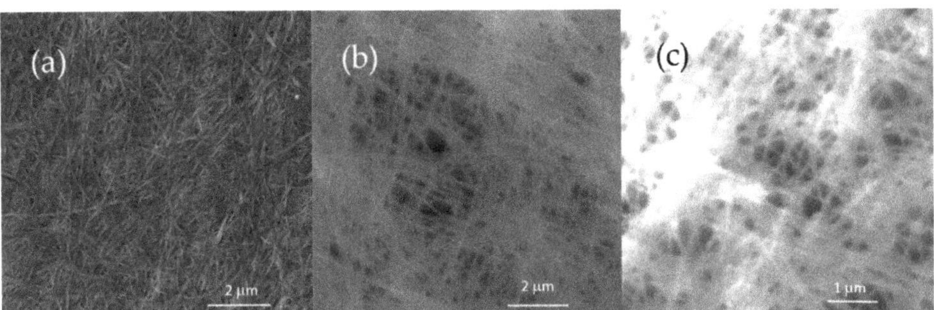

Figure 5. Nanostructure images of BC samples analyzed by SEM: (**a**) BC-AD, (**b**) BC-L transparent, (**c**) BC-L white.

It has been demonstrated that PM can change the material surface morphology, especially the surface roughness and porosity. This can lead to a material with different surface properties such as hydrophilicity [58]. The SEM analysis revealed ablation occurring, which is caused by surface etching during PM (Figure 6). The cellulose fibers appeared to be rugged and thinner after PM. Observed surface-terminated pores were bigger in the case of BC-AD. Those pores were wider and deeper with higher PM exposure time. The pores of BC-AD 240 s merged, forming a "brush" from the remaining parts of the fibers on surface. This very different structure is in agreement with the fact that the BC-AD 240 s sample was modified to the highest degree of the compared samples based on the gravimetric analysis. This is caused by the higher exposure time (compared to BC-AD 60 s) and by the fact that fibres in the air-dried sample are closer to each other than in the lyophilized ones. These results are in agreement with those published by Pertile et al. [44].

Figure 6. BC before and after plasmatic modification as analyzed by SEM: (**a**) BC-AD, (**b**) BC-AD 60 s, (**c**) BC-AD 240 s, (**d**) BC-L, (**e**) BC-L 60 s, (**f**) BC-L 240 s.

3.5. Gas Sorption Analysis of Porosity and Specific Surface Area

Specific surface area of unmodified samples and samples with 240 s PM were measured by gas sorption method. These results (Table 3) are in good agreement with the SEM observations. Firstly, compared to BC-AD, BC-L has almost five times higher porosity and specific surface area, because it maintains its structure (pellicle thickness) during the drying process. Further increase of pore volume and surface area was obtained after PM. This is due to PM fraying of the cellulose fibres. The number and size of the pores increases dramatically as well. Based on SEM images, we assumed the increase of porosity after PM for BC AD would be greater because of the "brush" structure, which was confirmed by the BJH analysis—the porosity increased almost 13 times for BC-AD, and 6 times for BC-L. Thus, after PM the specific surface area of the BC-L and BC-AD reached comparable values. On the basis of these results, we can conclude that PM in combination with the drying process leads to the increase of pores and surface area, which results in significant changes in surface morphology which can improve cell adhesion.

Table 3. Specific surface area (S_{BET}) and total pore volume (V_p) of BC samples measured by BET analysis.

Sample	S_{BET} $[m^2 \cdot g^{-1}]$	V_p $[cm^3 \cdot g^{-1}]$
BC-AD	9.9 ± 1.6	0.011 ± 0.002
BC-AD 240 s	140.5 ± 4.8	0.142 ± 0.008
BC-L	45.0 ± 0.7	0.056 ± 0.001
BC-L 240 s	156.3 ± 1.5	0.308 ± 0.015

3.6. Contact Angle and Hydrophilicity

The new morphology and higher oxygen content after PM can lead to changes in hydrophilicity of the material surface that can be represented by the water contact angle [59].

The contact angle measurement results (Table 4) showed that BC-AD is more hydrophobic than BC-L. The contact angles for those samples are $(63.91 \pm 2.69)°$ and $(34.74 \pm 6.8)°$, respectively. These differences can be attributed to the different specific surface area, which is smaller for BC-AD. For hydrophilic materials (contact angle less than $90°$) the higher porosity leads to a lower contact angle. This is due to the water drop being absorbed into the pores of a hydrophilic material. After PM, we observed further decrease of the contact angle that was greater in the case of BC-AD; both samples showed similar values after PM. This is consistent with much higher porosity increase for plasma-modified BC-AD compared to BC-L (Table 3). Additionally, an increase of surface oxygen content after PM represented by the O/C in XPS measurements (Table 2) leads to a more hydrophilic material. For comparison, Kurniawan et al. [45] also observed decrease of contact angle after PM with N_2 and O_2 plasmas while for CF_4 plasma there was an increase. However, Pertile et al. [44] noticed an increase of contact angle after N_2 PM. It is useful to note that each research group used different plasma discharge parameters.

Table 4. Values of water contact angle of BC samples measured by goniometry.

Sample	Contact Angle [°]
BC-AD	63.91 ± 2.69
BC-AD 60 s	25.42 ± 2.86
BC-AD 240 s	32.79 ± 2.00
BC-L	34.74 ± 6.80
BC-L 60 s	27.00 ± 2.80
BC-L 240 s	20.90 ± 1.90

3.7. In Vitro Tests of Cell Cultivation on BC Samples

For in vitro tests, we chose the BC-AD and BC-L samples, unmodified and after 240 s PM, in order to evaluate the effect of air drying, lyophilization, and PM on the colonization of the samples with human HaCaT keratinocytes.

On day 1 after seeding, the HaCaT cells on BC adhered generally in higher numbers than on the control PS wells, which was, however, more pronounced in BC-L than in BC-AD (Figure 7a). This result can be attributed to a higher porosity and a larger specific surface area of BC-L, which therefore provided more space for the initial cell attachment than BC-AD. Nevertheless, the initial adhesion of cells on BC-AD was significantly improved by PM, i.e., a technique which is generally used to enhance the attractiveness of various materials for cell adhesion. The main underlying mechanism of this improvement is increase in the material hydrophilicity, manifested by a significant decrease of water drop contact angle (from approx. 64° to 25°–33°; Table 4). On wettable materials, the cell adhesion-mediating proteins, such as fibronectin and vitronectin, spontaneously adsorb to the materials from biological fluids (including cell culture media), are attached in an active, physiological conformation, and are well-accessible for cell adhesion receptors (e.g., integrins) on cells [10,59,60]. BC-L was sufficiently wettable even before PM (contact angle of approx. 35°), and thus the PM did not further increase significantly the number of initially adhered cells, as observed on BC-AD samples. Moreover, the BC-AD samples modified with plasma were the first substrates on which the HaCaT cells started to form well-apparent and distinct islands typical for keratinocytes, which are important initial structures for creating a continuous cell layer (Figure 8).

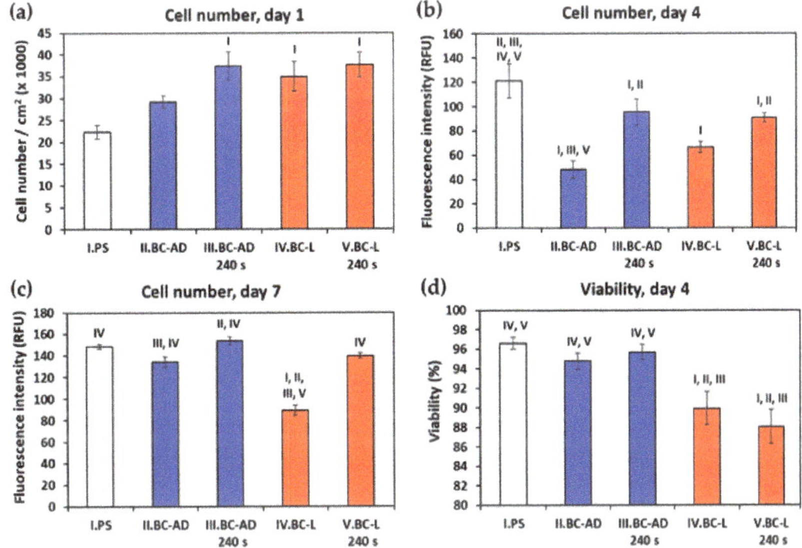

Figure 7. Number (**a–c**) and viability (**d**) of human HaCaT keratinocytes on days 1, 4 and 7 after seeding on control polystyrene wells (PS) or on BC. The data were obtained by (**a**) counting cells on 18–23 microscopic images, (**b,c**) measuring the intensity of fluorescence of cells stained with Hoechst 33258 on 7–20 microscopic images, or (**d**) trypan blue-exclusion test from four parallel samples of each experimental group. Mean ± S.E.M. (Standard Error of Mean), One way ANOVA, Student-Newman-Keuls Method. Statistically significant differences among the experimental groups ($p \leq 0.05$) are indicated above the columns by numbers of differing groups.

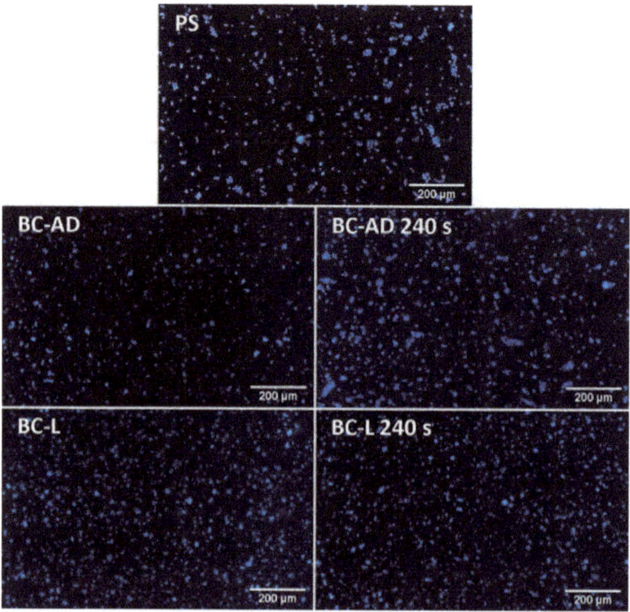

Figure 8. Human HaCaT keratinocytes on day 1 after seeding on control polystyrene wells (PS) or on BC. The cell nuclei were stained with Hoechst 33258. Olympus IX 51 epifluorescence microscope, DP 70 digital camera, obj. 4×, scale bar 200 μm.

On day 4 after seeding, however, the cell number became significantly lower on all tested BC samples in comparison with the reference PS wells (Figure 7b). In accordance with this, the cells on the images taken on day 4 covered a considerable part of the PS surface and were almost confluent, while the cells on the BC-AD and BC-L samples without PM were in an early phase of islet formation (Figure 9). This result can be explained by the fact that the flat PS surface provided a better support for the cell spreading (which is a prerequisite of the subsequent cell proliferation) than the rougher and more irregular BC surfaces. It was particularly apparent on BC-AD samples, which showed more rugged surface on SEM images (Figure 5). It is known from studies on osteoblasts that the increased material surface roughness often hampered proliferation of these cells [61,62]. This phenomenon could be even more pronounced in keratinocytes, which are epithelial cells with polarization (i.e., functional specialization) of their basal and apical cytoplasmic membrane, designated to cover surfaces of various organs, i.e., to live and grow in a 2D-like environment. In our earlier study, the negative effect of increased surface roughness on the cell proliferation was more apparent in endothelial cells, i.e., another type of epithelial-like cells, than in osteoblasts [63].

Nevertheless, the growth of HaCaT cells on BC-AD and BC-L was markedly improved by PM. On plasma-modified BC samples, the cells reached significantly higher cell numbers than on the unmodified samples (Figure 7b), and the cell islands on these samples developed into large cell colonies (Figure 9).

As revealed by a trypan blue exclusion test performed on day 4, the cells on the tested samples generally showed a high viability, ranging from approx. 88% to 97% (Figure 7d). Surprisingly, the lowest viability values were observed in cells cultivated on lyophilized BC, especially on samples modified with plasma. These samples contained the highest amount of oxygen (Table 2), which might be associated with damage to cells by reactive oxygen species. In addition, the surfaces of lyophilized samples are highly porous and contain numerous surface irregularities and areas of variable overall thickness and hardness, which can hamper cell spreading. It has been previously reported that keratinocytes prefer softer materials over harder ones [64,65]. Similarly, as mentioned above, the more rugged surface, which can be tolerated or even preferred by osteoblasts [66], can decrease the adhesion and proliferation of keratinocytes as well as other cells of soft tissues.

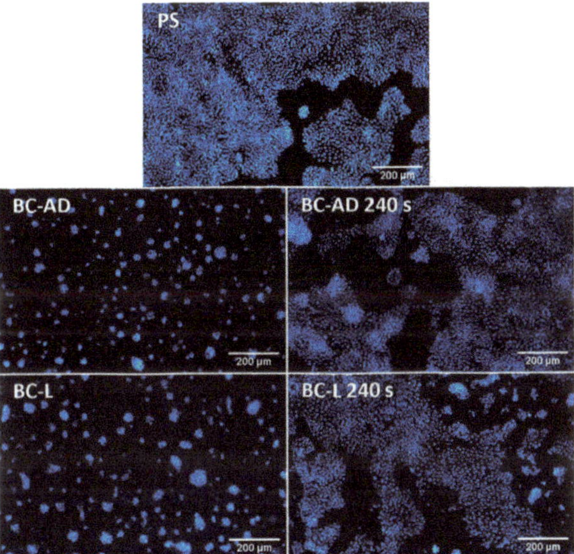

Figure 9. Human HaCaT keratinocytes on day 4 after seeding on control polystyrene wells (PS) or on BC. The cell nuclei were stained with Hoechst 33258. Olympus IX 51 epifluorescence microscope, DP 70 digital camera, obj. 4×, scale bar 200 µm.

On day 7 after seeding, the cell number on the tested BC samples equalized with that on reference PS; i.e., the initial increase in cell number on BC samples, observed on day 1, was lost (Figure 7c). On BC-L samples, the final cell number was even significantly lower than on the other samples. This can be a consequence of the decreased cell viability observed on day 4 (Figure 7d), due to some negative effects of BC-L samples on cell spreading and proliferation, such as their surface irregularities and possible presence of oxygen radicals. Additionally, higher porosity (Table 3) of BC-L materials can lead to higher swelling of these materials in water-containing environments, such as cell culture media, which can have rather negative effect on the cell adhesion, because it can decrease the toughness (rigidity) of the substrate material. It is known that very soft and deformable materials, such as hydrogels, cannot resist the traction forces generated by cells during their spreading and support the viable growth of cells [67,68].

Nevertheless, the cell colonization of all BC samples on day 7 significantly improved in comparison with the results from the 4th day of cultivation. From the images of cells on day 7 (Figure 10) it is evident that on PS and the plasma-modified BC, the cells are fully confluent, including those on BC-L, and in some parts of these samples, they created multilayer structures. On unmodified BC samples, the cells were still growing in colonies without reaching confluence, but these colonies have markedly enlarged in comparison with day 4.

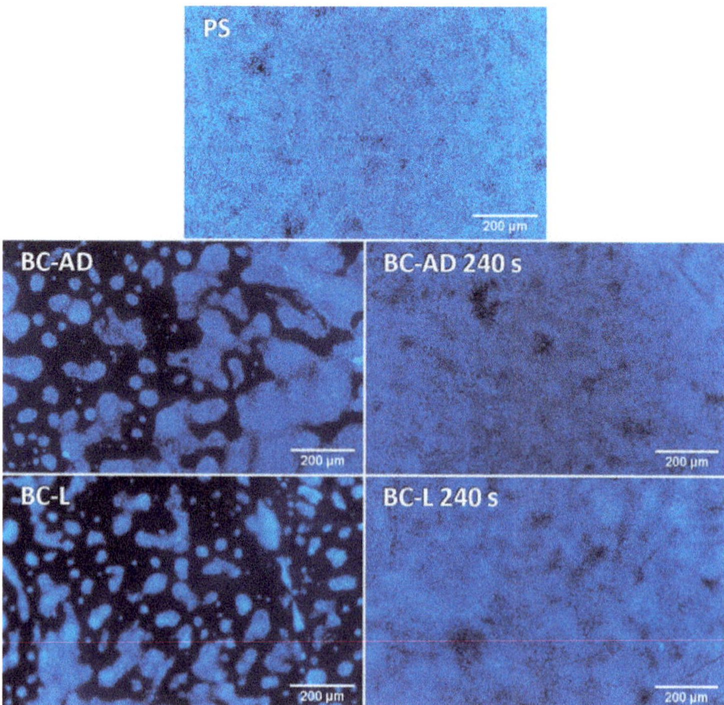

Figure 10. Human HaCaT keratinocytes on day 7 after seeding on control polystyrene wells (PS) or on BC. The cell nuclei were stained with Hoechst 33258. Olympus IX 51 epifluorescence microscope, DP 70 digital camera, obj. 4×, scale bar 200 µm.

Taken together, BC samples investigated in this study provided a good support for the adhesion, growth and viability of human HaCaT keratinocytes, comparable with standard tissue culture polystyrene, which is considered to be one of the most suitable materials for cell colonization. The effect of air drying and lyophilization on the cell colonization of BC was in general also comparable; it can be only distinguished that lyophilization had a positive effect on initial cell attachment, while the subsequent growth of cells was better on air-dried BC. However, the highest positive effect on the colonization of BC with cells has been provided by PM, as evident from the fact that the plasma-modified BC samples, either air-dried or lyophilized, were the only BC samples on which the cells were able to develop a continuous, confluent layer (an even multilayer) after one week of cultivation.

4. Conclusions

In our work, we described the differences between air drying and lyophilization with subsequent plasmatic modification of bacterial nanocellulose produced by the *Komagataeibacter sucrofermentans* bacteria strain (ATCC 700178). We found that BC-L materials maintained their structure, leading to higher porosity and specific surface area. Since the structure of BC-AD collapsed, the SEM revealed these samples to have a more rugged surface leading to almost two times higher contact angle. After plasmatic modification, the prepared samples showed a decrease of contact angle and increase of porosity and specific surface area. The O/C atomic ratio of the samples increased after PM, which suggests binding of the atmospheric oxygen to the activated surface immediately after modification. Compared to the literature, we obtained different values and trends of contact angles, morphology changes, and surface elemental composition. This is likely caused by the different parameters of the employed plasma discharge. Furthemore, we studied keratinocytes adhesion on our samples. We tested the initial adhesion (day 1) and subsequent growth (up to day 7) of human HaCaT keratinocytes on unmodified and plasma-modified BC-AD and BC-L. Due to its increased porosity and specific surface area, BC-L increased the initial adhesion of keratinocytes, but the subsequent cell growth was better on BC-AD. Modification with plasma for 240 s markedly accelerated the formation of typical keratinocyte islands and of continuous cell layers on both BC-AD and BC-L, which were comparable to those on standard cell culture polystyrene. Thus, the plasma-modified BC holds a great promise for skin tissue engineering and wound healing.

Author Contributions: Conceptualization, O.K. and L.B.; methodology, A.K. and L.S.; validation, O.K., L.B., and V.Š.; formal analysis, A.K., L.S., and O.K.; investigation, A.K., L.S., K.V., B.V., D.F., Z.K., and A.B.; resources, L.B. and V.Š.; data curation, A.K. and L.S.; writing—original draft preparation, A.K. and L.S.; writing—review and editing, O.K. and L.B.; visualization, A.K. and L.S.; supervision, O.K., L.B. and V.Š.; project administration, L.B. and V.Š.; funding acquisition, L.B. All authors have read and agreed to the published version of the manuscript.

Funding: This research was funded by GACR, grant number 20-01641S and OP VVV Project NANOTECH ITI II. No. CZ.02.1.01/0.0/0.0/18_069/0010045.

Conflicts of Interest: The authors declare no conflict of interest. The funders had no role in the design of the study; in the collection, analyses, or interpretation of data; in the writing of the manuscript, or in the decision to publish the results.

References

1. Jayabalan, R.; Malbaša, R.V.; Lončar, E.S.; Vitas, J.S.; Sathishkumar, M. A Review on Kombucha Tea-Microbiology, Composition, Fermentation, Beneficial Effects, Toxicity, and Tea Fungus. *Compr. Rev. Food Sci. Food Saf.* **2014**, *13*, 538–550. [CrossRef]
2. Brown, A.J. XLIII.—On an acetic ferment which forms cellulose. *J. Chem. Soc. Trans.* **1886**, *49*, 432–439. [CrossRef]
3. Matsutani, M.; Ito, K.; Azuma, Y.; Ogino, H.; Shirai, M.; Yakushi, T.; Matsushita, K. Adaptive mutation related to cellulose producibility in Komagataeibacter medellinensis (*Gluconacetobacter xylinus*) NBRC 3288. *Appl. Microbiol. Biotechnol.* **2015**, *99*, 7229–7240. [CrossRef] [PubMed]
4. Jonas, R.; Farah, L.F. Production and application of microbial cellulose. *Polym. Degrad. Stab.* **1998**, *59*, 101–106. [CrossRef]
5. Yamada, Y.; Yukphan, P.; Vu, H.T.L.; Muramatsu, Y.; Ochaikul, D.; Tanasupawat, S.; Nakagawa, Y. Description of *Komagataeibacter* gen. nov., with proposals of new combinations (*Acetobacteraceae*). *J. Gen. Appl. Microbiol.* **2012**, *58*, 397–404. [CrossRef] [PubMed]
6. Klemm, D.; Kramer, F.; Moritz, S.; Lindström, T.; Ankerfors, M.; Gray, D.; Dorris, A. Nanocelluloses: A New Family of Nature-Based Materials. *Angew. Chem. Int. Ed.* **2011**, *50*, 5438–5466. [CrossRef] [PubMed]
7. Younesi, M.; Akkus, A.; Akkus, O. Microbially-derived nanofibrous cellulose polymer for connective tissue regeneration. *Mater. Sci. Eng. C* **2019**, *99*, 96–102. [CrossRef]
8. Zeng, M.; Laromaine, A.; Roig, A. Bacterial cellulose films: Influence of bacterial strain and drying route on film properties. *Cellulose* **2014**, *21*, 4455–4469. [CrossRef]
9. Chen, S.-Q.; Cao, X.; Li, Z.; Zhu, J.; Li, L. Effect of lyophilization on the bacterial cellulose produced by different Komagataeibacter strains to adsorb epicatechin. *Carbohydr. Polym.* **2020**, *246*, 116632. [CrossRef]
10. Bacakova, L.; Pajorova, J.; Bacakova, M.; Skogberg, A.; Kallio, P.; Kolarova, K.; Svorcik, V. Versatile Application of Nanocellulose: From Industry to Skin Tissue Engineering and Wound Healing. *Nanomaterials* **2019**, *9*, 164. [CrossRef]
11. Nakagaito, A.; Yano, H. Novel high-strength biocomposites based on microfibrillated cellulose having nano-order-unit web-like network structure. *Appl. Phys. A* **2005**, *80*, 155–159. [CrossRef]
12. Czaja, W.; Krystynowicz, A.; Bielecki, S.; Brown, R.M. Microbial cellulose—The natural power to heal wounds. *Biomaterials* **2006**, *27*, 145–151. [CrossRef] [PubMed]
13. Zielińska, D.; Rydzkowski, T.; Thakur, V.K.; Borysiak, S. Enzymatic engineering of nanometric cellulose for sustainable polypropylene nanocomposites. *Ind. Crop. Prod.* **2021**, *161*, 113188. [CrossRef]
14. Yamanaka, S.; Watanabe, K.; Kitamura, N.; Iguchi, M.; Mitsuhashi, S.; Nishi, Y.; Uryu, M. The structure and mechanical properties of sheets prepared from bacterial cellulose. *J. Mater. Sci.* **1989**, *24*, 3141–3145. [CrossRef]
15. Trache, D.; Thakur, V.; Boukherroub, R. Cellulose Nanocrystals/Graphene Hybrids—A Promising New Class of Materials for Advanced Applications. *Nanomaterials* **2020**, *10*, 1523. [CrossRef]
16. Husemann, V.E.; Werner, R. Cellulose synthesis by *Acetobacter xylinum*. I. The molecular weight of bacterial cellulose and molecular weight distribution during the synthesis. *Die Makromol. Chem.* **1963**, *59*, 43–60. [CrossRef]
17. Rathinamoorthy, R.; Kiruba, T. Bacterial cellulose-A potential material for sustainable eco-friendly fashion products. *J. Nat. Fibers* **2020**, 1–13. [CrossRef]
18. Buruaga-Ramiro, C.; Valenzuela, S.V.; Valls, C.; Roncero, M.B.; Pastor, F.J.; Díaz, P.; Martinez, J. Development of an antimicrobial bioactive paper made from bacterial cellulose. *Int. J. Biol. Macromol.* **2020**, *158*, 587–594. [CrossRef]

19. Gallegos, A.M.A.; Carrera, S.H.; Parra, R.; Keshavarz, T.; Iqbal, H.M.N. Bacterial Cellulose: A Sustainable Source to Develop Value-Added Products—A Review. *BioResources* **2016**, *11*, 5641–5655. [CrossRef]

20. Uryu, M.; Kurihara, N. Acoustic Diaphragm and Method for Producing Same. U.S. Patent US5274199A, 20 April 1993.

21. Liu, X.; Souzandeh, H.; Zheng, Y.; Xie, Y.; Zhong, W.-H.; Wang, C. Soy protein isolate/bacterial cellulose composite membranes for high efficiency particulate air filtration. *Compos. Sci. Technol.* **2017**, *138*, 124–133. [CrossRef]

22. Hasan, N.; Biak, D.R.A.; Kamarudin, S. Application of Bacterial Cellulose (BC) in Natural Facial Scrub. *Int. J. Adv. Sci. Eng. Inf. Technol.* **2012**, *2*, 272–275. [CrossRef]

23. Picheth, G.F.; Pirich, C.; Sierakowski, M.R.; Woehl, M.A.; Sakakibara, C.N.; de Souza, C.F.; Martin, A.A.; da Silva, R.; de Freitas, R.A. Bacterial cellulose in biomedical applications: A review. *Int. J. Biol. Macromol.* **2017**, *104*, 97–106. [CrossRef] [PubMed]

24. Bodhibukkana, C.; Srichana, T.; Kaewnopparat, S.; Tangthong, N.; Bouking, P.; Martin, G.P.; Suedee, R. Composite membrane of bacterially-derived cellulose and molecularly imprinted polymer for use as a transdermal enantioselective controlled-release system of racemic propranolol. *J. Control. Release* **2006**, *113*, 43–56. [CrossRef]

25. Trovatti, E.; Freire, C.; Pinto, P.; Almeida, I.; da Costa, P.J.C.; Silvestre, A.; Neto, C.; Rosado, C. Bacterial cellulose membranes applied in topical and transdermal delivery of lidocaine hydrochloride and ibuprofen: In vitro diffusion studies. *Int. J. Pharm.* **2012**, *435*, 83–87. [CrossRef] [PubMed]

26. Cacicedo, M.L.; León, I.E.; Gonzalez, J.S.; Porto, L.M.; Álvarez, V.; Castro, G.R. Modified bacterial cellulose scaffolds for localized doxorubicin release in human colorectal HT-29 cells. *Colloids Surf. B Biointerfaces* **2016**, *140*, 421–429. [CrossRef]

27. Wang, J.; Gao, C.; Zhang, Y.; Wan, Y. Preparation and in vitro characterization of BC/PVA hydrogel composite for its potential use as artificial cornea biomaterial. *Mater. Sci. Eng. C* **2010**, *30*, 214–218. [CrossRef]

28. Jia, H.; Jia, Y.; Wang, J.; Hu, Y.; Zhang, Y.; Jia, S. Potentiality of Bacterial Cellulose as the Scaffold of Tissue Engineering of Cornea. In Proceedings of the 2009 2nd International Conference on Biomedical Engineering and Informatics, Tianjin, China, 17–19 October 2009; pp. 1–5.

29. Klemm, D.; Schumann, D.; Udhardt, U.; Marsch, S. Bacterial synthesized cellulose—Artificial blood vessels for microsurgery. *Prog. Polym. Sci.* **2001**, *26*, 1561–1603. [CrossRef]

30. Zang, S.; Zhang, R.; Chen, H.; Lu, Y.; Zhou, J.; Chang, X.; Qiu, G.; Wu, Z.; Yang, G. Investigation on artificial blood vessels prepared from bacterial cellulose. *Mater. Sci. Eng. C* **2015**, *46*, 111–117. [CrossRef] [PubMed]

31. Torgbo, S.; Sukyai, P. Bacterial cellulose-based scaffold materials for bone tissue engineering. *Appl. Mater. Today* **2018**, *11*, 34–49. [CrossRef]

32. Xu, C.; Ma, X.; Chen, S.; Tao, M.; Yuan, L.; Jing, Y. Bacterial Cellulose Membranes Used as Artificial Substitutes for Dural Defection in Rabbits. *Int. J. Mol. Sci.* **2014**, *15*, 10855–10867. [CrossRef] [PubMed]

33. Zhong, C. Industrial-Scale Production and Applications of Bacterial Cellulose. *Front. Bioeng. Biotechnol.* **2020**, *8*, 605374–605393. [CrossRef]

34. Bionext®. Available online: http://www.bennetthealth.net/Bionext/#Clinical (accessed on 22 April 2021).

35. Aung, B.J. Does A New Cellulose Dressing Have Potential in Chronic Wounds? *Podiatry Today* **2004**, *17*, 20–26.

36. Axcelon®. Available online: https://axcelonbp.com/nanoderm-ag/ (accessed on 22 April 2021).

37. Nexfill®. Available online: https://nexfill.com.br/ (accessed on 22 April 2021).

38. Sulaeva, I.; Henniges, U.; Rosenau, T.; Potthast, A. Bacterial cellulose as a material for wound treatment: Properties and modifications. A review. *Biotechnol. Adv.* **2015**, *33*, 1547–1571. [CrossRef] [PubMed]

39. Gorgieva, S. Bacterial Cellulose as a Versatile Platform for Research and Development of Biomedical Materials. *Processes* **2020**, *8*, 624. [CrossRef]

40. Helenius, G.; Bäckdahl, H.; Bodin, A.; Nannmark, U.; Gatenholm, P.; Risberg, B. In vivo biocompatibility of bacterial cellulose. *J. Biomed. Mater. Res. Part A* **2005**, *76*, 431–438. [CrossRef]

41. Torres, F.G.; Commeaux, S.; Troncoso, O.P. Biocompatibility of Bacterial Cellulose Based Biomaterials. *J. Funct. Biomater.* **2012**, *3*, 864–878. [CrossRef]

42. Chu, P.K.; Chen, J.Y.; Wang, L.P.; Huang, N. Plasma-surface modification of biomaterials. *Mater. Sci. Eng. R Rep.* **2002**, *36*, 143–206. [CrossRef]

43. Vosmanska, V.; Kolarova, K.; Rimpelova, S.; Svorcik, V. Surface modification of oxidized cellulose haemostat by argon plasma treatment. *Cellulose* **2014**, *21*, 2445–2456. [CrossRef]

44. Pertile, R.; Andrade, F.K.; Alves, C.; Gama, M. Surface modification of bacterial cellulose by nitrogen-containing plasma for improved interaction with cells. *Carbohydr. Polym.* **2010**, *82*, 692–698. [CrossRef]

45. Kurniawan, H.; Lai, J.-T.; Wang, M.-J. Biofunctionalized bacterial cellulose membranes by cold plasmas. *Cellulose* **2012**, *19*, 1975–1988. [CrossRef]

46. Vasconcellos, V.M.; Farinas, C.S. The effect of the drying process on the properties of bacterial cellulose films from *Gluconacetobacter hansenii*. *Chem. Eng. Trans.* **2018**, *64*, 145–150. [CrossRef]

47. Illa, M.P.; Sharma, C.S.; Khandelwal, M. Tuning the physiochemical properties of bacterial cellulose: Effect of drying conditions. *J. Mater. Sci.* **2019**, *54*, 12024–12035. [CrossRef]

48. Hestrin, S.; Schramm, M. Synthesis of cellulose by *Acetobacter xylinum*. 2. Preparation of freeze-dried cells capable of polymerizing glucose to cellulose. *Biochem. J.* **1954**, *58*, 345–352. [CrossRef]

49. Hsieh, J.-T.; Wang, M.-J.; Lai, J.-T.; Liu, H.-S. A novel static cultivation of bacterial cellulose production by intermittent feeding strategy. *J. Taiwan Inst. Chem. Eng.* **2016**, *63*, 46–51. [CrossRef]
50. Kim, S.-Y.; Kim, J.-N.; Wee, Y.-J.; Park, D.-H.; Ryu, H.-W. Production of Bacterial Cellulose by *Gluconacetobacter* sp. RKY5 Isolated from Persimmon Vinegar. *Appl. Biochem. Biotechnol.* **2006**, *131*, 705–715. [CrossRef]
51. Bae, S.; Sugano, Y.; Shoda, M. Improvement of bacterial cellulose production by addition of agar in a jar fermentor. *J. Biosci. Bioeng.* **2004**, *97*, 33–38. [CrossRef]
52. Wee, Y.-J. Isolation and characterization of a bacterial cellulose-producing bacterium derived from the persimmon vinegar. *Afr. J. Biotechnol.* **2011**, *10*, 16267–16276. [CrossRef]
53. Švorčík, V.; Kolářová, K.; Slepička, P.; Mackova, A.; Novotná, M.; Hnatowicz, V. Modification of surface properties of high and low density polyethylene by Ar plasma discharge. *Polym. Degrad. Stab.* **2006**, *91*, 1219–1225. [CrossRef]
54. Makarem, M.; Lee, C.M.; Kafle, K.; Huang, S.; Chae, I.; Yang, H.; Kubicki, J.D.; Kim, S.H. Probing cellulose structures with vibrational spectroscopy. *Cellulose* **2019**, *26*, 35–79. [CrossRef]
55. Hofstetter, K.; Hinterstoisser, B.; Salmén, L. Moisture uptake in native cellulose—The roles of different hydrogen bonds: A dynamic FT-IR study using Deuterium exchange. *Cellulose* **2006**, *13*, 131–145. [CrossRef]
56. Fan, M.; Dai, D.; Huang, B. Fourier Transform Infrared Spectroscopy for Natural Fibres. In *Fourier Transform-Materials Analysis*, 1st ed.; Salih, S., Ed.; InTech: Rijeka, Croatia, 2012; pp. 45–68.
57. Maréchal, Y.; Chanzy, H. The hydrogen bond network in I β cellulose as observed by infrared spectrometry. *J. Mol. Struct.* **2000**, *523*, 183–196. [CrossRef]
58. Yang, J.; Bei, J.; Wang, S. Enhanced cell affinity of poly (D,L-lactide) by combining plasma treatment with collagen anchorage. *Biomaterials* **2002**, *23*, 2607–2614. [CrossRef]
59. Slepička, P.; Trostová, S.; Kasálková, N.S.; Kolská, Z.; Sajdl, P.; Švorčík, V. Surface Modification of Biopolymers by Argon Plasma and Thermal Treatment. *Plasma Process. Polym.* **2011**, *9*, 197–206. [CrossRef]
60. Ku, S.H.; Ryu, J.; Hong, S.K.; Lee, H.; Park, C.B. General functionalization route for cell adhesion on non-wetting surfaces. *Biomaterials* **2010**, *31*, 2535–2541. [CrossRef]
61. Steinerova, M.; Matejka, R.; Stepanovska, J.; Filova, E.; Stankova, L.; Rysova, M.; Martinova, L.; Dragounova, H.; Domonkos, M.; Artemenko, A.; et al. Human osteoblast-like SAOS-2 cells on submicron-scale fibers coated with nanocrystalline diamond films. *Mater. Sci. Eng. C* **2021**, *121*, 111792. [CrossRef]
62. Vandrovcova, M.; Tolde, Z.; Vanek, P.; Nehasil, V.; Doubková, M.; Trávníčková, M.; Drahokoupil, J.; Buixaderas, E.; Borodavka, F.; Novakova, J.; et al. Beta-Titanium Alloy Covered by Ferroelectric Coating–Physicochemical Properties and Human Osteoblast-Like Cell Response. *Coatings* **2021**, *11*, 210. [CrossRef]
63. Grausova, L.; Kromka, A.; Bacakova, L.; Potocký, S.; Vanecek, M.; Lisa, V. Bone and vascular endothelial cells in cultures on nanocrystalline diamond films. *Diam. Relat. Mater.* **2008**, *17*, 1405–1409. [CrossRef]
64. Bacakova, M.; Pajorova, J.; Stranska, D.; Hadraba, D.; Lopot, F.; Riedel, T.; Brynda, E.; Zaloudkova, M.; Bacakova, L. Protein nanocoatings on synthetic polymeric nanofibrous membranes designed as carriers for skin cells. *Int. J. Nanomed.* **2017**, *12*, 1143–1160. [CrossRef]
65. Bacakova, M.; Pajorova, J.; Broz, A.; Hadraba, D.; Lopot, F.; Zavadakova, A.; Vistejnova, L.; Beno, M.; Kostic, I.; Jencova, V.; et al. A two-layer skin construct consisting of a collagen hydrogel reinforced by a fibrin-coated polylactide nanofibrous membrane. *Int. J. Nanomed.* **2019**, *14*, 5033–5050. [CrossRef]
66. Stankova, L.; Fraczek-Szczypta, A.; Blazewicz, M.; Filova, E.; Blazewicz, S.; Lisa, V.; Bacakova, L. Human osteoblast-like MG 63 cells on polysulfone modified with carbon nanotubes or carbon nanohorns. *Carbon* **2014**, *67*, 578–591. [CrossRef]
67. Engler, A.; Sen, S.; Sweeney, H.L.; Discher, D.E. Matrix Elasticity Directs Stem Cell Lineage Specification. *Cell* **2006**, *126*, 677–689. [CrossRef] [PubMed]
68. Bacakova, L.; Filova, E.; Parizek, M.; Ruml, T.; Svorcik, V. Modulation of cell adhesion, proliferation and differentiation on materials designed for body implants. *Biotechnol. Adv.* **2011**, *29*, 739–767. [CrossRef] [PubMed]

 *nanomaterials*

Article

Antibacterial Multi-Layered Nanocellulose-Based Patches Loaded with Dexpanthenol for Wound Healing Applications

Daniela F. S. Fonseca [1], João P. F. Carvalho [1], Verónica Bastos [2], Helena Oliveira [2], Catarina Moreirinha [1], Adelaide Almeida [2], Armando J. D. Silvestre [1], Carla Vilela [1,*] and Carmen S. R. Freire [1,*]

[1] CICECO–Aveiro Institute of Materials, Department of Chemistry, University of Aveiro, 3810-193 Aveiro, Portugal; danielafonseca@ua.pt (D.F.S.F.); joao.pedro.carvalho@ua.pt (J.P.F.C.); catarina.fm@ua.pt (C.M.); armsil@ua.pt (A.J.D.S.)

[2] Department of Biology and CESAM, University of Aveiro, 3810-193 Aveiro, Portugal; veronicabastos@ua.pt (V.B.); holiveira@ua.pt (H.O.); aalmeida@ua.pt (A.A.)

* Correspondence: cvilela@ua.pt (C.V.); cfreire@ua.pt (C.S.R.F.)

Received: 6 November 2020; Accepted: 8 December 2020; Published: 9 December 2020

Abstract: Antibacterial multi-layered patches composed of an oxidized bacterial cellulose (OBC) membrane loaded with dexpanthenol (DEX) and coated with several chitosan (CH) and alginate (ALG) layers were fabricated by spin-assisted layer-by-layer (LbL) assembly. Four patches with a distinct number of layers (5, 11, 17, and 21) were prepared. These nanostructured multi-layered patches reveal a thermal stability up to 200 °C, high mechanical performance (Young's modulus ≥4 GPa), and good moisture-uptake capacity (240–250%). Moreover, they inhibited the growth of the skin pathogen *Staphylococcus aureus* (3.2–log CFU mL^{-1} reduction) and were non-cytotoxic to human keratinocytes (HaCaT cells). The in vitro release profile of DEX was prolonged with the increasing number of layers, and the time-dependent data imply a diffusion/swelling-controlled drug release mechanism. In addition, the in vitro wound healing assay demonstrated a good cell migration capacity, headed to a complete gap closure after 24 h. These results certify the potential of these multi-layered polysaccharides-based patches toward their application in wound healing.

Keywords: oxidized bacterial cellulose; chitosan; alginate; layer-by-layer assembly; multi-layered patches; dexpanthenol; wound healing

---

1. Introduction

Natural polymers are being explored as building blocks to engineer multifunctional sustainable materials with sophisticated architectures via biomacromolecular assembly strategies. Among the prevailing methodologies to design such materials, the layer-by-layer (LbL) assembly is a simple, versatile and well-established technique with high relevance in the drug delivery domain [1]. In fact, the LbL assembly, which entails the sequential adsorption of complementary species (mostly based on opposing charges) on a substrate, is enabling the creation of drug reservoirs either in the form of planar films or capsules with a high-level of control of drug administration, targeted delivery, and lower side effects. The presence of multiple layers in a single system facilitates the design of materials with a profusion of bioactive functionalities (e.g., antioxidant activity, antibacterial activity, and wound healing capacity) that are typical to their individual precursors [1].

Polysaccharides are amid the most commonly utilized components for the design of multi-layered systems applied on drug delivery, despite the constraint of not yielding robust and free-standing films or capsules without the use of a template, either sacrificial or non-sacrificial (e.g., glass slides, mesoporous silica, or calcium carbonate nanoparticles) [1]. In the context of planar films, bacterial cellulose (BC) membranes, viz. an exopolysaccharide produced by some non-pathogenic bacteria [2,3], can be used simultaneously as a non-sacrificial template and a drug reservoir. In fact, this robust and high purity cellulosic substrate, with a web-like entangled morphology, possesses in situ moldability that allows its direct biosynthesis in the form of membranes with personalized size and shape, revealing good in vitro biocompatibility and in vivo skin compatibility [4], alongside high water retention capacity, nanostructured porous network, high mechanical performance, and tailorable surface chemistry [5]. This set of intrinsic and unique features has been explored for the development of BC-based cutaneous drug delivery systems loaded with both hydrophilic or lipophilic active pharmaceutical ingredients (APIs) [6,7]. Nevertheless, the majority of the studies involves the simple diffusion of an API solution into the BC network followed by drying, which is a methodology that hinders the API selection and, in most instances, does not enable a controlled and targeted delivery of the APIs, particularly when they are highly water-soluble [7,8]. Herein, the use of the LbL assembly methodology, which can be performed in mild conditions in most substrates and enables the modulation of the composition, stability, and surface functionality of those substrates [1], can be a major asset for engineering BC-based membranes for wound healing applications, where a controlled and targeted drug delivery is an essential requirement.

Our interest in developing multi-layered BC derived patches by LbL assembly for cutaneous drug delivery originates from the idea of combining the peculiar properties of a freestanding exopolysaccharide robust membrane (i.e., BC) with the properties of hydrophilic ionic polysaccharides (i.e., chitosan (CH) and alginate (ALG)), and an API with wound healing potential (i.e., dexpanthenol (DEX) used in several pharmaceutical formulations in the field of dermatology and skin care). Although BC (either in its pure or oxidized forms) has already been individually blended with CH/ALG [9,10] or DEX [11], and BC nanocrystals have been used as a filler in an alginate matrix that was then covered with CH and gelatin polyelectrolytes via LbL assembly [12], there are no studies—as far as our literature search could verify—dedicated to the development of multi-layered patches composed of oxidized BC (OBC), CH, and ALG via spin-assisted LbL assembly technology, where the nanostructured porous cellulosic substrate plays the simultaneous role of a non-sacrificial template and a drug reservoir.

In this perspective, the goal of the present study is to design multi-layered patches composed of three polysaccharides, namely BC, CH, and ALG, as antibacterial cutaneous drug delivery vehicles for DEX, viz. a water-soluble API with wound healing ability. The multi-layered patches with a stratified structure were assembled via spin-assisted LbL methodology using a DEX-loaded OBC membrane as the initial negatively charged substrate, followed by the alternate adsorption of positively charged CH and negatively charged ALG polyelectrolytes, in a total of 5, 11, 17, and 21 layers. An in-depth analysis of the structure, morphology, thermal stability, mechanical properties, and moisture-uptake capacity of the multi-layered patches is presented, as well as the in vitro drug release profile at physiological pH, the in vitro antibacterial activity against *Staphylococcus aureus*, the in vitro cytotoxicity toward human HaCaT keratinocyte cells and the wound healing potential.

2. Materials and Methods

2.1. Chemicals, Bacteria, and Cells

Dexpanthenol (DEX, pharmaceutical secondary standard), glycerol (≥99.5%), phosphate buffered saline (PBS, pH 7.4), sodium hypochlorite solution (available chlorine 10–15%), sodium bromide (≥99%), sodium alginate (ALG, alginic acid sodium salt from brown algae, medium viscosity ≥2000 cP (2%, 25 °C)), 2,2,6,6-tetramethylpiperidine-1-oxyl (TEMPO free radical, >98%), tryptic soy broth (TSB), tryptic soy agar (TSA), 3-(4,5-dimethylthiazolyl-2)-2,5-diphenyltetrazolium bromide (MTT, 98%), and dimethyl sulfoxide (DMSO, ≥99.9%) were supplied by Sigma-Aldrich (Sintra, Portugal). Dulbecco's Modified Eagle's Medium (DMEM), fetal bovine serum (FBS), L-glutamine, penicillin/streptomycin, and fungizone were purchased from Gibco® (Life Technologies, Carlsbad, CA, USA). Type 1 ultrapure water (resistivity of 18.2 MΩ cm (25 °C)) was filtered by a Simplicity® Water Purification System (Merck Millipore, Darmstadt, Germany). Other chemicals and solvents were of laboratory grade.

Chitosan (CH from shrimp, degree of deacetylation of 98%, viscosity of 2900 cP (1% solution in 1% of acetic acid)) was obtained from Mahtani Chitosan Pvt. Ltd. India. For purification, CH was dissolved in a 1% (*v/v*) aqueous acetic acid solution, filtered, and precipitated by neutralization with NaOH 0.5 M up to a pH of 8.5. The precipitate was rinsed with water until neutral pH, followed by freeze-drying [13].

Bacterial cellulose (BC) was biosynthesized in our laboratory by the *Gluconacetobacter sacchari* bacterial strain in the form of wet membranes [14]. *Staphylococcus aureus* (ATCC 6538) bacterium was received from American Type Culture Collection (ATCC) collection (Manassas, VA, USA). The nontumorigenic immortalized human keratinocyte HaCaT cell line was acquired from Cell Lines Services (Eppelheim, Germany).

2.2. Preparation of Oxidized Bacterial Cellulose via TEMPO-Mediated Oxidation

The oxidized bacterial cellulose (OBC) membranes were prepared via TEMPO-mediated oxidation as described elsewhere, with some modifications [15]. Shortly, 50 g of never-dried BC were suspended into deionized water containing TEMPO (0.13 g) and NaBr (1.3 g) under magnetic stirring. Then, 0.267 g of NaClO aqueous solution was slowly added, and the pH of the solution was maintained at 10.5 by adding a 0.5 M NaOH aqueous solution. The temperature was kept at 30 °C throughout the reaction for 30 min. The oxidation was stopped by quenching with 40 mL of absolute ethanol, followed by washing with deionized water. The OBC membranes (carboxyl content of 0.54 ± 0.09 mmol g^{-1}) were stored in distilled water at 4 °C until further use.

2.3. Preparation of Multi-Layered Patches Loaded with Dexpanthenol

First, ALG and CH solutions (1 mg mL^{-1}) were prepared in ultrapure water and 1% (*v/v*) acid acetic aqueous solution, respectively. Wet OBC membranes (6×4 cm^2) were weighted, drained, and then soaked in 5 mL of an aqueous buffered solution (pH 7.4) of DEX (0.15% *w/v*) with glycerol (1% *w/v*). The membranes were stirred at 100 rpm and 30 °C for 1 h to allow the complete absorption of the solution (viz. 100% entrapment efficiency) and then left to dry at 30 °C for 24 h. The DEX-loaded OBC patch was coated with 250 μL of CH (zeta (ζ)-potential: 57 ± 3 mV (Zetasizer Nano ZS, Malvern Panalytical, Cambridge, United Kingdom)) and 250 μL of ALG (ζ-potential: -40 ± 3 mV) aqueous solutions by spin coating (Spin 150, APT GmbH–Automation und Produktionstechnik, Germany) at 2000 rpm during 15 s. A total of 5, 11, 17, or 21 layers of the polysaccharides were deposited, starting and finishing with CH, as compiled in Table 1.

Table 1. List of patches with the respective dry weight and thickness values.

Patch [a]	Layers	Dry weight/mg [b]	Thickness/μm [b]
OBC	OBC	267 ± 10	45 ± 10
OBC/DEX	OBC(DEX)	275 ± 11	50 ± 15
OBC/DEX/CH/ALG_5	OBC(DEX)-CH-ALG-CH-ALG-CH	277 ± 13	55 ± 10
OBC/DEX/CH/ALG_11	OBC(DEX)-CH-ALG-CH-ALG-CH-ALG-CH-ALG-CH-ALG-CH	279 ± 13	61 ± 11
OBC/DEX/CH/ALG_17	OBC(DEX)-CH-ALG-CH-ALG-CH-ALG-CH-ALG-CH-ALG-CH-ALG-CH-ALG-CH-ALG-CH	284 ± 10	65 ± 16
OBC/DEX/CH/ALG_21	OBC(DEX)-CH-ALG-CH-ALG-CH-ALG-CH-ALG-CH-ALG-CH-ALG-CH-ALG-CH-ALG-CH-ALG-CH-ALG-CH	286 ± 11	69 ± 13

[a] All patches contain glycerol as a plasticizer (1% *w/v*, 2.1 mg per cm^2 of patch) that was incorporated inside the OBC; [b] Values are expressed as mean ± standard deviation.

2.4. Characterization Techniques

The thickness of the specimens was measured with a Mitutoyo coolant-proof digimatic micrometer MDC-25PX (Mitutoyo Corporation, Tokyo, Japan). All measurements were executed at fifteen different locations of the specimens, randomly selected, with an accuracy of 1 µm.

Attenuated total reflection-Fourier transform infrared (ATR-FTIR) spectra were computed on a Perkin-Elmer FT-IR System Spectrum BX spectrophotometer (Perkin-Elmer Inc., Waltham, MA, USA) coupled with a single horizontal Golden Gate ATR cell (Specac®, London, UK), in the wavenumber range of 600–4000 cm^{-1} at a resolution of 4 cm^{-1} over 32 scans.

Scanning electron micrographs (SEM) were recorded in a HR-FESEM SU-70 Hitachi microscope (Hitachi High-Technologies Corporation, Tokyo, Japan). The specimens were placed on an aluminum plate and formerly coated with a carbon film.

Thermogravimetric analysis (TGA) was undertaken in a SETSYS SETARAM TGA analyzer (SETARAM Instrumentation, Lyon, France). The specimens were heated from 25 °C to 800 °C at a heating rate of 10 °C min^{-1} under a nitrogen atmosphere.

Tensile mechanical tests were conducted on a uniaxial Instron 5966 testing apparatus (Instron Corp., Rockville, MD, USA) in traction mode at a deformation rate of 10 mm min^{-1} using a 500 N static load cell and gauge length of 30 mm. The specimens were rectangular strips (5 × 1 cm^2) dried at 40 °C. All measurements were held on seven replicates.

The moisture-uptake capacity was assessed by laying the dry specimens (2 × 2 cm^2) in a conditioned cabinet with 100% relative humidity at 25 °C for 22 h. After that period, the weight (W_w) of the moist specimens was measured, and the moisture-uptake was determined according to the equation:

$$\text{Moisture} - \text{uptake} \ (\%) = (W_w - W_0) \times W_0^{-1} \times 100 \tag{1}$$

where W_0 is the initial weight of the dry specimens [16].

2.5. In Vitro Drug Release Assays

DEX-loaded patches (2 × 2 cm^2) were immersed in a vessel containing 30 mL of 0.01 M PBS (pH 7.4) and the dissolution tests were then carried out at 32 °C and 50 rpm during 90 h. At regular time intervals, aliquots of 3 mL of each solution were withdrawn, and the same volume of fresh PBS was added to maintain a constant volume. The DEX concentration in each aliquot was determined by UV–Vis spectroscopy (Shimadzu UV-1800 UV–Vis spectrophotometer, Shimadzu Corp., Kyoto, Japan) at 206 nm with the linear calibration curve: $y = 0.135x + 0.0167$ ($R^2 = 0.9978$), in the range of 10–100 µm mL^{-1}. The DEX content at each time was plotted as a cumulative percentage release concentration calculated using the formula:

$$C_{\text{cumulative}} = C_n + [(3 \times C_{n-1})/30] \tag{2}$$

where C_n and C_{n-1} are the concentrations of DEX at time n and $n - 1$, respectively. Nine replicates were performed for each specimen [8].

2.6. In Vitro Antibacterial Activity

The antibacterial activity of the OBC-based membranes was examined against *S. aureus*. The bacterial pre-inoculum cultures were grown for 24 h in TSB growth medium at 37 °C under horizontal shaking at 120 rpm until reaching a concentration of 10^8 to 10^9 colony forming units per mL (CFU mL^{-1}). Each film specimen (4 × 6 cm^2) was placed in contact with 5 mL of bacterial suspension via a ten-fold dilution of the overnight grown culture in PBS (pH 7.4). A bacterial cell suspension was tested as the control while OBC membranes, neat and DEX-loaded, were tested as blank references. All specimens were incubated at 37 °C under horizontal shaking at 120 rpm. At 24 h contact time, aliquots (500 µL) of each specimen and controls were collected and the bacterial concentration

(CFU mL^{-1}) was determined by plating serial dilutions on TSA medium. The plates were incubated at 37 °C for 48 h. The CFU were determined on the most appropriate dilution on the agar plates. Two independent experiments were carried out and, for each, two replicates were plated. The bacteria log reduction of the specimens was calculated as follows:

$$\log CFU \text{ mL}^{-1} reduction = \log CFU_{control} - \log CFU_{patch} \tag{3}$$

2.7. In Vitro Cytotoxicity Assay

The cytotoxicity of the patches was evaluated in human keratinocytes cell line (HaCaT cells) by the MTT assay [17]. Briefly, cells were grown in complete DMEM medium supplemented with 10% FBS, 2 mM L-glutamine, 10,000 U mL^{-1} penicillin/streptomycin, and 250 µg mL^{-1} fungizone at 37 °C in 5% CO_2 humidified atmosphere. Cells were observed daily under an inverted phase-contrast Eclipse TS100 microscope (Nikon, Tokyo, Japan). The tests were performed in triplicates of OBC, OBC/DEX, and OBC/DEX/CH/ALG_21. Specimens of 1×1 cm^2 were prepared, sterilized by ultraviolet radiation, and then incubated with 1.0 mL of complete DMEM medium at 37 °C with 5% CO_2 for 24 h to prepare the specimen extract.

Meanwhile, one 96-well plate was prepared with 5×4 wells filled up with 6000 cells/well and the cells were then incubated in complete culture medium for 24 h for adhesion. After that time, cell culture medium (in the 96-well plates) was substituted by 100 µL of each of the extracts obtained from the incubated specimens and cells were then additionally incubated for either 24 h or 48 h. As a negative control, HaCaT cells were incubated with complete DMEM medium and treated identically as described for the specimens.

At the end of the incubation time, 50 µL of MTT (at a concentration of 1 g L^{-1}) was added to each well and incubated for 4 h at 37 °C in a 5% CO_2 humidified atmosphere. After that, the culture medium with MTT was removed and replaced by 150 µL of DMSO and the plate was positioned in a shaker for 2 h in the dark to totally dissolve the formazan crystals. The absorbance of the specimens was measured with a BioTek Synergy HT plate reader (Synergy HT Multi-Mode, BioTeK, Winooski, VT, USA) at 570 nm with blank corrections. The cell viability was calculated with respect to the control cells:

$$\text{Cell viability } (\%) = \left[\left(Abs_{sample} - Abs_{DMSO} \right) / \left(Abs_{control} - Abs_{DMSO} \right) \right] \times 100 \tag{4}$$

where the Abs_{sample} is the absorbance of the specimen; Abs_{DMSO} is the absorbance of the $DMSO$ solvent; and $Abs_{control}$ is the absorbance of the control.

2.8. In Vitro Wound Healing (Scratch) Assay

The effect of the multi-layered patches on the migration capability of HaCaT keratinocyte cells was evaluated using the scratch assay. Cells were seeded in a 6-well plate at 5×10^5 cells/well and the linear wound was generated with a sterile 200 µL plastic pipette tip. Specimens of 1×1 cm^2 were prepared, sterilized by ultraviolet radiation, and then incubated with 1.0 mL of complete DMEM medium at 37 °C, with 5% CO_2 for 24 h to prepare the specimen extract. The tests were performed in triplicates of OBC/DEX, OBC/DEX/CH/ALG_11, and OBC/DEX/CH/ALG_21. As a negative control, HaCaT cells were treated identically as described for the specimens. Each specimen extract was applied to the scratch and the cells were then further incubated for 9 and 24 h at 37 °C, with 5% CO_2. Optical micrographs (Eclipse TS100 microscope, Nikon, Tokyo, Japan) of the scratched area of each condition were taken and compared to the corresponding 0 h time point.

2.9. Statistical Analysis

The statistical analysis was conducted with OriginPro 9.0.0 (OriginLab Corporation, Northampton, MA, USA), where the statistical significance was assessed by the analysis of variance (ANOVA) and Tukey's test with the statistical significance established at $p < 0.05$.

3. Results and Discussion

Polysaccharides-based multi-layered patches composed of OBC, CH, and ALG were fabricated by spin-assisted LbL coating assembly technology, as illustrated in Figure 1. This assembly methodology was selected due to its simplicity and the stratified structure it conveys to the materials [1]. Therefore, the first step to prepare the multi-layered membranes comprised the functionalization of BC via the environmentally friendly TEMPO-mediated oxidation to obtain oxidized bacterial cellulose membranes with negatively charged groups at the surface (Figure 1a). The water-soluble radical reagent TEMPO is a catalytic oxidant that jointly with NaBr and NaClO selectively converts the cellulose C6 hydroxyl groups into carboxylic moieties in aqueous media, without compromising the morphological integrity of the fibers [18].

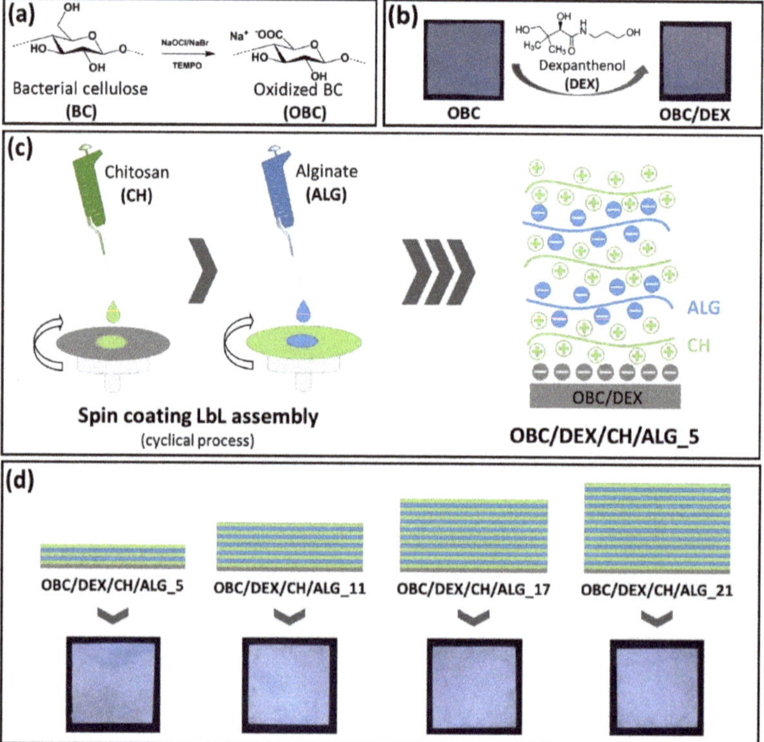

Figure 1. Schematic representation of (**a**) TEMPO-mediated oxidation of BC yielding OBC, (**b**) loading of OBC with DEX, (**c**) spin coating LbL assembly of the polysaccharides multi-layered patch with 5 layers, and (**d**) digital photographs of the OBC/DEX/CH/ALG patches with 5, 11, 17, and 21 layers.

The OBC membrane was then loaded with DEX (i.e., OBC/DEX membrane, Figure 1b), which is a water-soluble active pharmaceutical ingredient (API) used in several pharmaceutical formulations in the field of dermatology and skin care due to its capacity to act as a skin moisturizer, skin barrier restorer, and facilitator of wound healing [19]. All membranes contain 0.32 mg of DEX per cm^2 of OBC, selected based on equivalent commercial formulations [19]. Further to this, glycerol (1% *w/v*, 2.1 mg per cm^2 of membrane) was incorporated in the membrane as a plasticizer to upsurge their malleability and conformability to the skin surface [20].

Afterward, four multi-layered patches with a distinct number of layers (5, 11, 17, and 21) of the positively charged CH (ζ-potential: 57 ± 3 mV) and the negatively charged ALG (ζ-potential:

−40 ± 3 mV) polyelectrolytes were prepared (Figure 1c). CH and ALG polysaccharides were selected as the oppositely charged polyelectrolytes because of their efficient film-forming ability and vast applicability in the LbL assembly technology [1]. Furthermore, CH was chosen as the final layer in the four multi-layered membranes due to its known antibacterial activity toward a wide variety of microorganisms [13,21], largely linked to the presence of amino groups [22]. This bioactive property is particularly pertinent in the context of skin regeneration to avoid the growth of pathogenic microorganisms that will contribute to microbial wound infections [23,24].

All multi-layered patches are whitish and homogeneous, as displayed in the photographs of Figure 1d, confirming the even distribution of DEX and of the polyelectrolytes (i.e., CH and ALG). Predictably, the weight and thickness of the multi-layered patches increased with the increasing number of layers as listed Table 1. The nanostructured patches were characterized in terms of structure by ATR-FTIR spectroscopy, morphology by SEM, thermal stability by TGA, mechanical properties by tensile tests and moisture-uptake capacity. Additionally, the in vitro drug release profile, in vitro antibacterial activity, in vitro cytotoxicity, and in vitro wound healing potential were also evaluated.

3.1. Structural and Morphological Characterization

The modification of BC via TEMPO-mediated oxidation to obtain OBC with negatively charged groups was confirmed by ATR-FTIR spectroscopy through the presence of the typical absorption bands of the BC backbone at 1029 cm^{-1} (C–O stretching), 1316 cm^{-1} (O–H in plane bending), 2897 cm^{-1} (C–H stretching), and 3340 cm^{-1} (O–H stretching) [25], and the emergence of the absorption band at 1590–1640 cm^{-1} assigned to the antisymmetric vibration of ionized carboxyl groups (ν_{as}(COO$^-$), Figure 2a) [26]. Moreover, the SEM micrograph of the OBC membrane surface (Figure 2c) shows the characteristic three-dimensional nanofibrillar structure of BC [5,14]. This means that the oxidation of BC did not affect the morphology (Figure 2c) nor the visual appearance of the membranes.

The incorporation of DEX via diffusion into the three-dimensional porous network of OBC was also validated by ATR-FTIR spectroscopy (Figure 2b) through the presence of the typical absorption bands of each component (Figure 2a), whilst the ones allocated to DEX, namely at 3308 cm^{-1} (O–H and N–H stretching), 1636 cm^{-1} (C=O stretching, amide I band), and 1535 cm^{-1} (amide II band) [26,27] are not completely observable given the small amount of API present in the patch (0.32 mg of DEX per cm^2 of OBC). In the SEM micrograph of the surface of the OBC/DEX patch (Figure 2c), the nanofibrillar structure of OBC is still visible, suggesting that all DEX was absorbed into the three-dimensional porous network of OBC and none remains at the surface of the patch, confirming the above vibrational spectroscopy data.

The ATR-FTIR spectra of the multi-layered patches (Figure 2b) portrayed similarities with those of their precursors (Figure 2a), mainly with OBC, but also with chitosan at 3342 cm^{-1} (O–H stretching), 1638 cm^{-1} (C=O stretching and N–H bending), 1592 cm^{-1} (–NH$_2$ bending), 1380 cm^{-1} (–CH$_2$ bending), 1148 cm^{-1} (C–O–C antisymmetric stretching), 1078 and 1032 cm^{-1} (C–O stretching) [28], and ALG at 3250 cm^{-1} (O–H stretching), and 1596 cm^{-1} (COO$^-$ antisymmetric stretching) and 1406 cm^{-1} (COO$^-$ symmetric stretching) [29]. Moreover, the absorption bands of glycerol at 1100 cm^{-1} and 1030 cm^{-1} (C–O stretching vibrations of alcohols) and 3250 cm^{-1} (O–H stretching), [7] are fully superimposed with the vibrations of the polysaccharides, viz. OBC, CH, and ALG.

Regarding the morphology (Figure 2c), the SEM micrographs of the surface of the multi-layered membranes did not disclose the three-dimensional fibrillar structure of OBC because it was completely covered by the various CH and ALG layers. Furthermore, the micrographs also depicted the formation of micron-size particles credited to the polyelectrolytes, which underlines the homogeneous and total coverage of the negatively charged OBC substrate. Indeed, in the case of the patch with 21 layers, the micrograph resembles the formation of a particulate film.

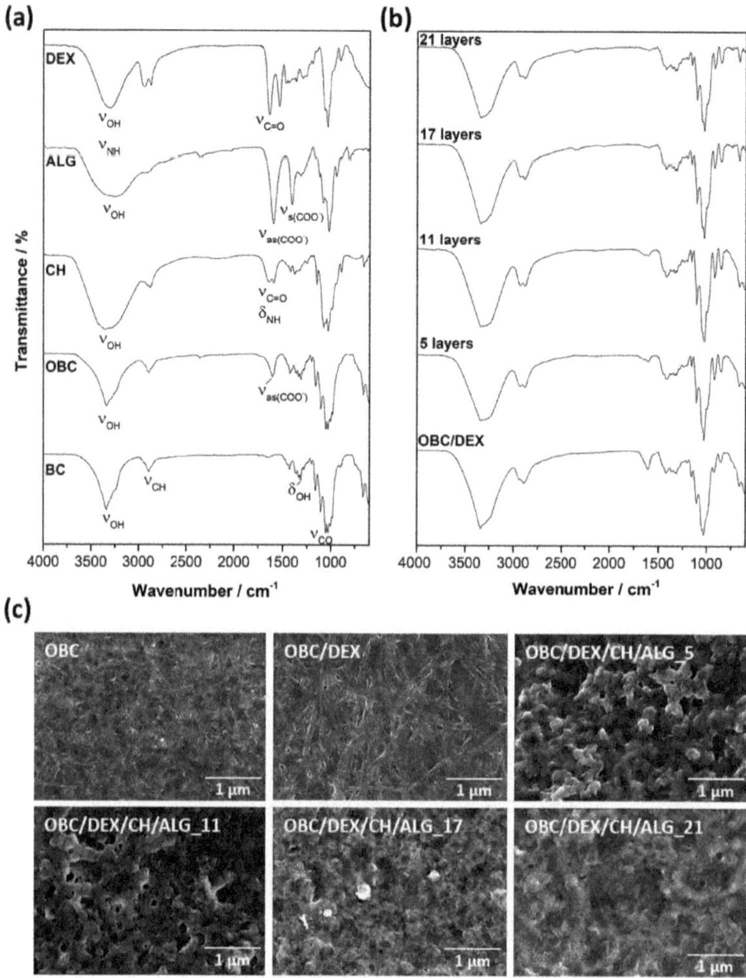

Figure 2. (**a**,**b**) ATR-FTIR spectra (vibrational modes: ν = stretching, δ = bending) of BC, OBC, CH, ALG, DEX (**a**), OBC/DEX, and OBC/DEX/CH/ALG-based multi-layered patches with 5, 11, 17, and 21 layers (**b**), and (**c**) SEM surface micrographs (×30.0 k magnification) of OBC, OBC/DEX, and OBC/DEX/CH/ALG-based multi-layered patches with 5, 11, 17, and 21 layers.

3.2. Thermal Stability and Mechanical Properties

The thermal stability of the multi-layered patches was measured by thermogravimetric analysis under a nitrogen atmosphere. According to Figure 3, the data did not vary appreciably with the increasing number of polyelectrolyte layers (i.e., 5, 11, 17, and 21 layers), with all membranes exhibiting a double weight-loss degradation profile with thermal stability up to about 200 °C (Figure 3b). Apart from the initial volatilization of water at ca. 100 °C, the small first step with a maximum rate of decomposition temperature at ca. 240 °C is allocated to the volatilization of glycerol [30] and degradation of DEX [27], while the second step with a maximum rate of decomposition temperature in the range 320–330 °C is related to the polysaccharide skeleton degradation, namely OBC [31], CH [13], and ALG [32]. In this second-step, the maximum degradation temperature slightly decreases with the increasing number of layers, most likely due to the marginally lower thermal stability of the polyelectrolytes [13,32]. Furthermore, the final residue of the multi-layered patches at 800 °C varied between 25% for OBC/DEX/CH/ALG_5 and 35% for OBC/DEX/CH/ALG_21, which is mainly attributed to the increase of CH and ALG in the membranes. However, it is worth remarking that all multi-layered OBC/DEX/CH/ALG-based membranes are thermally stable up to 200 °C, and therefore, can undergo the sterilization procedures (e.g., autoclaving at ca. 120 °C) required for biomedical applications.

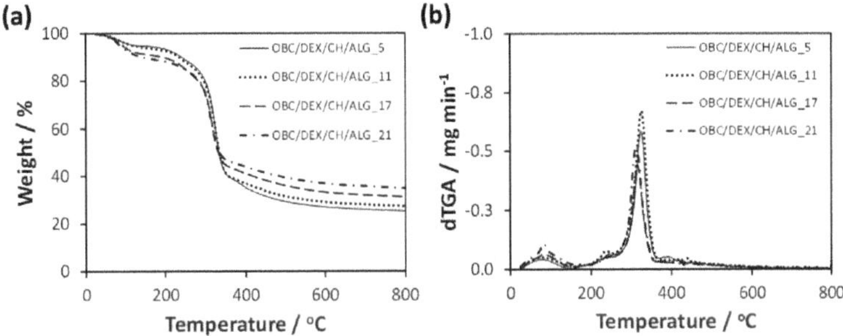

Figure 3. (a) Thermograms and (b) derivative curves of the multi-layered patches: OBC/DEX/CH/ALG_5, OBC/DEX/CH/ALG_11, OBC/DEX/CH/ALG_17, and OBC/DEX/CH/ALG_21.

The mechanical properties of the multi-layered polysaccharides-based patches were also analyzed, and the data obtained from the stress-strain curves are condensed in Table 2. The negatively charged OBC substrate exhibits an elongation at break of 3.4 ± 0.6%, tensile strength of 150 ± 25 MPa and Young's modulus of 8.2 ± 1.4 GPa. Furthermore, the inclusion of DEX (dissolved in a buffered solution with glycerol (1% w/v)) into the OBC porous structure promoted a decrease in both Young's modulus and tensile strength to 3.5 ± 1.2 GPa and 97 ± 15 MPa, respectively, complemented by an increase in the elongation at break to 4.9 ± 0.6%. This behavior is comparable to the trend obtained for pure BC membranes after the incorporation of APIs such as diclofenac [33] and lidocaine [34] into the BC three-dimensional structure. Furthermore, Tanrıverdi et al. also reported a decrease in the tensile strength of synthetic nanofiber mats, namely poly(lactic-*co*-glycolic acid), poly(ethylene oxide), and poly(ε-caprolactone), loaded with DEX [35].

Table 2. Young's modulus, tensile strength, and elongation at break of the OBC, OBC/DEX, and the multi-layered patches.

Patch [a]	Young's Modulus/GPa	Tensile Strength/MPa	Elongation at Break/%
OBC	8.2 ± 1.4	150 ± 25	3.4 ± 0.6
OBC/DEX	3.5 ± 1.2	97 ± 15	4.9 ± 0.6
OBC/DEX/CH/ALG_5	4.2 ± 0.8	142 ± 39	4.3 ± 0.7
OBC/DEX/CH/ALG_11	4.4 ± 0.8	154 ± 52	4.0 ± 1.1
OBC/DEX/CH/ALG_17	9.8 ± 2.7	166 ± 45	2.0 ± 0.4
OBC/DEX/CH/ALG_21	10.9 ± 4.0	188 ± 57	2.2 ± 0.4

[a] All patches contain glycerol as a plasticizer.

After the LbL assembly of the oppositely charged polyelectrolytes (i.e., CH and ALG) on the negatively charged OBC substrate, the mechanical performance of the multi-layered membranes increased with the increasing number of layers. The Young's modulus increased from 4.2 ± 0.8 GPa for OBC/DEX/CH/ALG_5 to 10.9 ± 4.0 GPa for OBC/DEX/CH/ALG_21, whereas the tensile strength increased from 142 ± 39 MPa for OBC/DEX/CH/ALG_5 to 188 ± 57 MPa for OBC/DEX/CH/ALG_21. On the contrary, the elongation at break decreased from 4.3 ± 0.7% for OBC/DEX/CH/ALG_5 to 2.2 ± 0.4% for OBC/DEX/CH/ALG_21, meaning that the membranes became less elastic as the number of layers increased from 5 to 21. Nonetheless, these hydrophilic multi-layered patches are satisfactorily malleable to conform to the surface of the skin.

Worth noting is the fact that these OBC/DEX/CH/ALG multi-layered patches exhibited a higher tensile strength than the similar non-multi-layered films based on hydrogen peroxide oxidized BC (HOBC) or periodic acid oxidized BC (POBC) containing CH and ALG, which were prepared by blending the homogenized HOBC or POBC pulp with a CH and ALG gel solution, and studied for application as wound dressings [10].

3.3. Moisture-Uptake Capacity

The moisture-uptake capacity of the multi-layered patches was determined to anticipate their interaction with moisture and exudates when applied to wound sites. Hence, the patches were left in a chamber with 100% relative humidity at 25 °C for 22 h. According to Figure 4, all OBC/DEX/CH/ALG patches absorbed moisture over time and were able to reach values of 240–250% after 22 h of exposure. Furthermore, the time-dependent data did not vary appreciably with the increasing number of layers (i.e., 5, 11, 17, and 21 layers), meaning that these patches present an analogous moisture-uptake capacity. This behavior is obviously credited to the hydrophilic nature of the precursors of the OBC/DEX/CH/ALG-based membranes, as in fact shown for the non-multi-layered films composed of OBC, CH, and ALG [10].

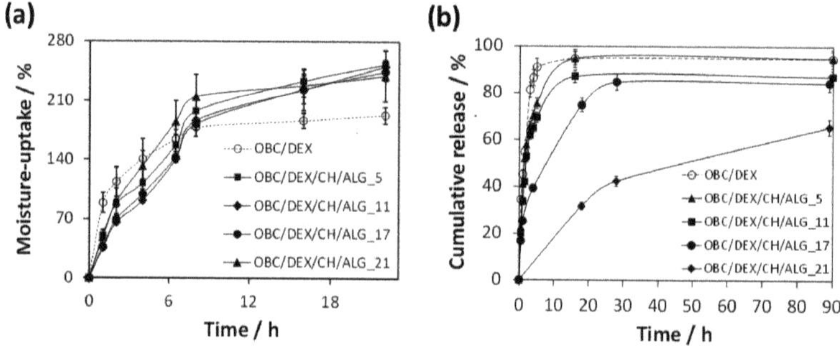

Figure 4. (**a**) Moisture-uptake capacity and (**b**) DEX cumulative release profile of OBC, OBC-loaded with DEX, and multi-layered polysaccharides-based patches.

Therefore, any of these multi-layered polysaccharide-based patches can promote (if necessary) a moist environment at the skin interface and remove the surplus of exudates or other body fluids when applied to the skin surface.

3.4. In Vitro Drug Release Profile

The in vitro release profile of the multi-layered polysaccharides-based patches was tested in PBS (pH 7.4) at 32 °C that simulates the usual pH and temperature of the skin [36]. Figure 4b shows the cumulative release profiles of DEX from the OBC/DEX/CH/ALG-based membranes, which were compared with the in vitro release of DEX from a membrane composed solely of OBC and DEX.

In general, the release profile of the patches is quite standard, with an initial burst followed by a plateau where the DEX release achieves the highest value. While the OBC/DEX and OBC/DEX/CH/ALG_5 patches reached a maximum cumulative release of ca. 95% after 16 h, the OBC/DEX/CH/ALG_11 achieved ca. 87% after 16 h and OBC/DEX/CH/ALG_17 attained ca. 85% after 28 h. In the case of the OBC/DEX/CH/ALG_21 patch, it only reached a maximum cumulative release of ca. 65% after 90 h. This is a direct indication that the multi-layered systems retarded the DEX release from the OBC membrane, particularly for 21 layers. Moreover, this slow-release profile is suited for wounds that require longer therapeutic treatments combined with a local skin moisturizer with antibacterial properties and wound healing ability.

When compared with non-multi-layered BC-based membranes loaded with water-soluble APIs [7,8], these OBC/DEX/CH/ALG multi-layered patches presented a controlled release profile that only reached high cumulative release values after several hours.

The in vitro drug release curves (Figure 4b) were further adjusted to kinetic models to identify the diffusional release mechanism of DEX from the multi-layered OBC/DEX/CH/ALG-based membranes. The data were better fitted to the Korsmeyer–Peppas kinetic model [37,38], $M_t/M_\infty = kt^n$, with M_t as the DEX content released at time t, M_∞ is the DEX content released at infinite time, k is the kinetic constant, and n is the diffusion constant denoting the release mechanism [37,39]. Based on this kinetic model, only the cumulative release values of $M_t/M_\infty < 60\%$ are fitted, and therefore, a release exponent (n) of 0.45 was obtained for OBC/DEX, 0.68 for OBC/DEX/CH/ALG_5, 0.69 for OBC/DEX/CH/ALG_11, 0.68 for OBC/DEX/CH/ALG_17, and 0.70 for OBC/DEX/CH/ALG_21 (Table 3). These curve fitting parameters are representative of Fickian diffusion (i.e., diffusion-controlled drug release mechanism, in the case of OBC/DEX, and non-Fickian transport ($0.5 < n < 1.0$) (i.e., diffusion and swelling controlled drug release mechanism), for the case of the multi-layered polysaccharide-based membranes [37–39].

Table 3. Fitting parameters of the Korsmeyer–Peppas kinetic model and the corresponding drug release mechanism.

Patch	Release Exponent (n)	Regression Coefficient (R^2)	Drug Release Mechanism
OBC/DEX	0.45	0.997	Fickian
OBC/DEX/CH/ALG_5	0.68	0.998	Non-Fickian
OBC/DEX/CH/ALG_11	0.69	0.997	Non-Fickian
OBC/DEX/CH/ALG_17	0.68	0.998	Non-Fickian
OBC/DEX/CH/ALG_21	0.70	0.999	Non-Fickian

3.5. In Vitro Antibacterial Activity

The antibacterial activity of the multi-layered polysaccharide-based patches was assayed against *S. aureus* (a Gram-positive bacterium) for 24 h, as shown in Figure 5a. This bacterium has been extensively studied because it is the culprit of several skin infections, particularly in patients with skin injuries [23]. As expected, neither the control nor the OBC membrane hindered the growth of the Gram-positive pathogenic microorganism. This observation is consistent with the literature, since OBC does not inhibit the growth of *S. aureus*, as previously reported [40]. Nevertheless, when DEX is loaded into the OBC membrane, the bacterial growth of *S. aureus* was reduced by 0.8–log CFU mL^{-1} after 24 h. Although DEX is not an antibacterial agent [41], it seems that its combination with OBC conveyed a small growth inhibition to the OBC/DEX membrane.

Regarding the multi-layered polysaccharide-based patches, they are responsible for ca. 3.2–log CFU mL^{-1} reduction after 24 h (Figure 5a). Since the general tendency of the four membranes was the same, it is fair to claim that all OBC/DEX/CH/ALG patches are antibacterial agents due to the presence of chitosan. In fact, several studies have shown that the marine polysaccharide derived from chitin (i.e., chitosan) presents antibacterial activity toward a wide variety of microorganisms [21] including *S. aureus* [13], largely linked to the presence of amino groups [22,24]. This bioactive property is of paramount importance for skin regeneration to avert the growth of pathogenic microorganisms causing microbial wound infections [23,24].

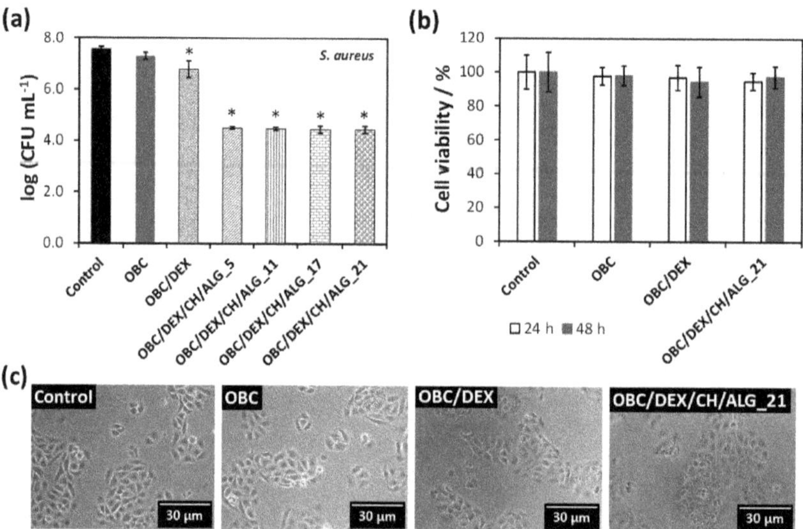

Figure 5. (**a**) Antibacterial activity (24 h of exposure) of OBC, OBC-loaded with DEX, and multi-layered patches (the symbol * represents the means with a significant difference ($p < 0.05$) from the control and OBC); (**b**) cell viability of HaCaT cells after 24 h and 48 h of exposure to negative control, OBC, OBC/DEX, and OBC/DEX/CH/ALG_21, and (**c**) the corresponding optical micrographs of HaCaT cells after 24 h of exposure.

3.6. In Vitro Cytotoxicity Assays

The cytotoxic behavior of the multi-layered patch with 21 layers (OBC/DEX/CH/ALG_21) as well as that of OBC and OBC/DEX was assessed in human HaCaT keratinocyte cells via the indirect MTT assay. This cell line was elected because it has been intensely used as a model for epidermal cells [42–45]. Figure 5b shows that the HaCaT cells metabolic activity was normal for the negative control (100% cell viability) as well as for the OBC membrane (ca. 98%) for both exposure times (24 and 48 h). Therefore, the OBC membrane is non-cytotoxic toward HaCaT cells, which is coherent with the results achieved in the literature for other cell lines such as NIH3T3 fibroblast cells [31]. Furthermore, the cell viability of the OBC patch loaded with DEX was almost unaltered with values of 97 ± 7% after 24 h and 94 ± 9% after 48 h. This behavior was predictable because DEX is a non-cytotoxic drug used in several pharmaceutical formulations in the field of dermatology and skin care [19].

With regard to the multi-layered patch with the higher number of layers (i.e., OBC/DEX/ CH/ALG_21), it was also deemed as a non-cytotoxic material because the cell viability was 95 ± 5% after 24 h and 97 ± 6% after 48 h (Figure 5b). Furthermore, the optical micrographs of the HaCaT cells (Figure 5c) support the cell viability data by evidencing that neither the cell morphology nor the cell count varied in relation to the control after 24 h of cell incubation with the OBC, OBC/DEX, and OBC/DEX/CH/ALG_21 membranes. Actually, all values were high above the 70% threshold of cell viability [46], which reflects the potential in vivo performance of these safe and compatible multi-layered patches for application in the biomedical field.

3.7. In Vitro Wound Healing Activity

Having proved that the multi-layered polysaccharide-based patches are robust, thermally stable, and present a controlled drug release profile, antibacterial activity against *S. aureus* and non-cytotoxicity toward HaCaT cells, the next step focused on the wound healing potential of these patches as delivery vehicles for DEX. Therefore, the wound healing activity of the multi-layered polysaccharide-based patches with 11 and 21 layers (OBC/DEX/CH/ALG_11 and OBC/DEX/CH/ALG_21) as well as of the OBC/DEX patch was assessed using the in vitro wound healing scratch assay [47]. Therefore, the influence of the patches on the cell migration was studied for HaCaT keratinocyte cells after 9 and 24 h of exposure. Predictably, and as shown in Figure 6, the cells cultivated in a medium containing only culture medium (i.e., control) presented a good migration capacity with the complete occlusion of the scratch after 24 h.

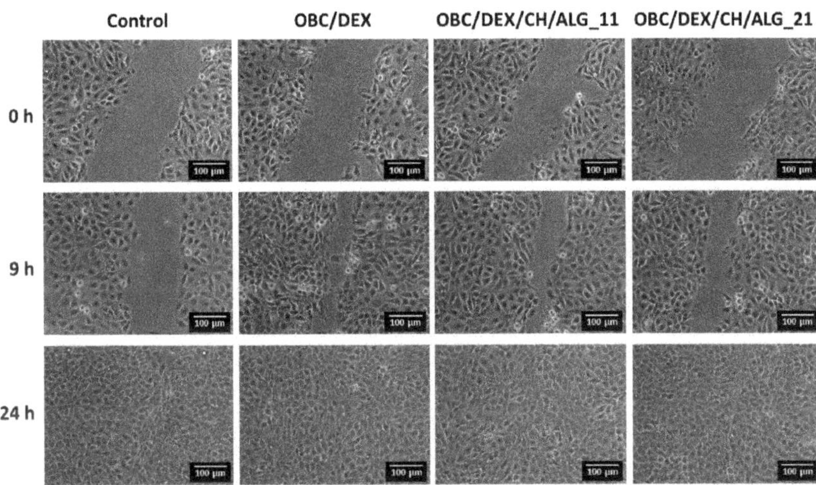

Figure 6. Optical micrographs of the scratch assay of the HaCaT cells after 9 and 24 h of exposure to control, OBC/DEX, OBC/DEX/CH/ALG_11, and OBC/DEX/CH/ALG_21 patches.

Overall, and in the case of the patches, the HaCaT keratinocyte cells situated near the boundary of the scratched region moved toward the vacant region as time progressed, achieving a full closure after 24 h. Furthermore, the cell density at the back of the scratch site is also growing with time, which points to the proliferation of the HaCaT keratinocyte cells. However, in terms of temporal progression, the patches containing DEX, namely OBC/DEX, OBC/DEX/CH/ALG_11 and OBC/DEX/CH/ALG_21, demonstrated higher coverage of the wound-like gap area after 9 h, induced by cell growth and migration, when compared to the positive control. Despite the analogous behavior of the HaCaT keratinocyte cells after 24 h of exposure to the three patches (Figure 6), the optical micrographs after 9 h show a slightly slower cell growth and migration for the patches with 11 and 21 layers. In fact, the behavior of the different patches after 9 h of exposure (Figure 6) seems to be in line with the in vitro drug release profile of DEX, where the OBC/DEX patch is the one releasing the higher amount of DEX in the shorter period of time and the multi-layered OBC/DEX/CH/ALG_21 membrane with a controlled drug release profile is the one releasing the lower content of DEX after 24 h, as depicted in Figure 4b. Similar observations in terms of wound closure as a function of time were documented by Sharma et al. [48] for curcumin conjugated with hyaluronic acid, which are two natural compounds known to expedite the wound healing process, and by Silva et al. [16] for multifunctional nanofibrous patches composed of nanocellulose and lysozyme nanofibers.

Given that wound closure is a signal of wound healing ability [49], these antibacterial and non-cytotoxic multi-layered polysaccharide-based patches have the potential to foster wound healing and, simultaneously, hinder skin infections. Moreover, the drug release profile can be tuned to the specific type of wound and required treatment time by simply adjusting the number of polysaccharide layers.

4. Conclusions

Functional multi-layered patches composed of OBC, CH, ALG, and DEX were prepared for wound healing applications. Four multi-layered patches with different number of layers, viz. 5, 11, 17, and 21, were assembled by the spin assisted LbL coating technique of oppositely charged CH and ALG polyelectrolytes onto the OBC cellulosic substrate loaded with DEX. These patches with a stratified structure had thermal stability up to almost 200 °C, mechanical properties with a Young's modulus higher than 4 GPa, and good moisture-uptake capacity (240–250%). Additionally, they inhibited the growth of *Staphylococcus aureus* with 3.2–log CFU mL^{-1} reduction and are non-cytotoxic to human keratinocytes (HaCaT cells) with a cell viability of about 97% after 48 h. The in vitro DEX release profile from the patches is time-dependent and is prolonged with the increasing number of layers, and the in vitro wound healing assay showed a good migration capacity with the full occlusion of the wound scratch after 24 h. These data validate the potentiality of these antibacterial multi-layered polysaccharides-based patches as vehicles for the topical delivery of DEX, which is a promoter of wound healing.

Author Contributions: Conceptualization, C.V. and C.S.R.F.; Investigation, D.F.S.F., J.P.F.C., V.B., C.M. and C.V.; Resources, H.O., A.A., A.J.D.S. and C.S.R.F.; Writing—original draft preparation C.V.; Writing—review and editing, D.F.S.F., J.P.F.C., V.B., H.O., C.M., A.A., A.J.D.S., C.V. and C.S.R.F.; Supervision, C.V. and C.S.R.F.; Funding acquisition, H.O., A.A., A.J.D.S. and C.S.R.F. All authors have read and agreed to the published version of the manuscript.

Funding: This work was developed within the scope of the projects CICECO-Aveiro Institute of Materials (UIDB/50011/2020 & UIDP/50011/2020) and CESAM (UIDB/50017/2020), financed by national funds through the Portuguese Foundation for Science and Technology (FCT)/MCTES. FCT is also acknowledged for the doctoral grant to D.F.S.F. (PD/BD/115621/2016), and the research contracts under Scientific Employment Stimulus to H.O. (CEECIND/04050/2017) and C.V. (CEECIND/00263/2018). The AgroForWealth project (CENTRO-01-0145-FEDER-000001) is acknowledged for the research grant to J.P.F.C., while the research contract of V.B. (CDL-CTTRI-161-ARH/2018) was funded by the FCT project POCI-01-0145-FEDER-031794.

Conflicts of Interest: The authors declare no conflict of interest.

References

1. Vilela, C.; Figueiredo, A.R.P.; Silvestre, A.J.D.; Freire, C.S.R. Multilayered materials based on biopolymers as drug delivery systems. *Expert Opin. Drug Deliv.* **2017**, *14*, 189–200. [CrossRef]

2. Wang, J.; Tavakoli, J.; Tang, Y. Bacterial cellulose production, properties and applications with different culture methods—A review. *Carbohydr. Polym.* **2019**, *219*, 63–76. [CrossRef]

3. Jacek, P.; Dourado, F.; Gama, M.; Bielecki, S. Molecular aspects of bacterial nanocellulose biosynthesis. *Microb. Biotechnol.* **2019**, *12*, 633–649. [CrossRef]

4. Almeida, I.F.; Pereira, T.; Silva, N.H.C.S.; Gomes, F.P.; Silvestre, A.J.D.; Freire, C.S.R.; Sousa Lobo, J.M.; Costa, P.C. Bacterial cellulose membranes as drug delivery systems: An in vivo skin compatibility study. *Eur. J. Pharm. Biopharm.* **2014**, *86*, 332–336. [CrossRef]

5. Klemm, D.; Cranston, E.D.; Fischer, D.; Gama, M.; Kedzior, S.A.; Kralisch, D.; Kramer, F.; Kondo, T.; Lindström, T.; Nietzsche, S.; et al. Nanocellulose as a natural source for groundbreaking applications in materials science: Today's state. *Mater. Today* **2018**, *21*, 720–748. [CrossRef]

6. Silvestre, A.J.D.; Freire, C.S.R.; Neto, C.P. Do bacterial cellulose membranes have potential in drug-delivery systems? *Expert Opin. Drug Deliv.* **2014**, *11*, 1113–1124. [CrossRef]

7. Silva, N.H.C.S.; Mota, J.P.; de Almeida, T.S.; Carvalho, J.P.F.; Silvestre, A.J.D.; Vilela, C.; Rosado, C.; Freire, C.S.R. Topical drug delivery systems based on bacterial nanocellulose: Accelerated stability testing. *Int. J. Mol. Sci.* **2020**, *21*, 1262. [CrossRef]

8. Carvalho, J.P.F.; Silva, A.C.Q.; Bastos, V.; Oliveira, H.; Pinto, R.J.B.; Silvestre, A.J.D.; Vilela, C.; Freire, C.S.R. Nanocellulose-based patches loaded with hyaluronic acid and diclofenac towards aphthous stomatitis treatment. *Nanomaterials* **2020**, *10*, 628. [CrossRef]

9. Wichai, S.; Chuysinuan, P.; Chaiarwut, S.; Ekabutr, P.; Supaphol, P. Development of bacterial cellulose/alginate/chitosan composites incorporating copper (II) sulfate as an antibacterial wound dressing. *J. Drug Deliv. Sci. Technol.* **2019**, *51*, 662–671. [CrossRef]

10. Chang, W.-S.; Chen, H.-H. Physical properties of bacterial cellulose composites for wound dressings. *Food Hydrocoll.* **2016**, *53*, 75–83. [CrossRef]

11. Haimer, E.; Wendland, M.; Schlufter, K.; Frankenfeld, K.; Miethe, P.; Potthast, A.; Rosenau, T.; Liebner, F. Loading of bacterial cellulose aerogels with bioactive compounds by antisolvent precipitation with supercritical carbon dioxide. *Macromol. Symp.* **2010**, *294*, 64–74. [CrossRef]

12. Yan, H.; Chen, X.; Feng, M.; Shi, Z.; Zhang, D.; Lin, Q. Layer-by-layer assembly of 3D alginate-chitosan-gelatin composite scaffold incorporating bacterial cellulose nanocrystals for bone tissue engineering. *Mater. Lett.* **2017**, *209*, 492–496. [CrossRef]

13. Vilela, C.; Pinto, R.J.B.; Coelho, J.; Domingues, M.R.M.; Daina, S.; Sadocco, P.; Santos, S.A.O.; Freire, C.S.R. Bioactive chitosan/ellagic acid films with UV-light protection for active food packaging. *Food Hydrocoll.* **2017**, *73*, 120–128. [CrossRef]

14. Vilela, C.; Silva, A.C.Q.; Domingues, E.M.; Gonçalves, G.; Martins, M.A.; Figueiredo, F.M.L.; Santos, S.A.O.; Freire, C.S.R. Conductive polysaccharides-based proton-exchange membranes for fuel cell applications: The case of bacterial cellulose and fucoidan. *Carbohydr. Polym.* **2020**, *230*, 115604. [CrossRef]

15. Saito, T.; Kimura, S.; Nishiyama, Y.; Isogai, A. Cellulose nanofibers prepared by TEMPO-mediated oxidation of native cellulose. *Biomacromolecules* **2007**, *8*, 2485–2491. [CrossRef]

16. Silva, N.H.C.S.; Garrido-Pascual, P.; Moreirinha, C.; Almeida, A.; Palomares, T.; Alonso-Varona, A.; Vilela, C.; Freire, C.S.R. Multifunctional nanofibrous patches composed of nanocellulose and lysozyme nanofibers for cutaneous wound healing. *Int. J. Biol. Macromol.* **2020**, *165*, 1198–1210. [CrossRef]

17. Mosmann, T. Rapid colorimetric assay for cellular growth and survival: Application to proliferation and cytotoxicity assays. *J. Immunol. Methods* **1983**, *65*, 55–63. [CrossRef]

18. Habibi, Y. Key advances in the chemical modification of nanocelluloses. *Chem. Soc. Rev.* **2014**, *43*, 1519–1542. [CrossRef]

19. Proksch, E.; De Bony, R.; Trapp, S.; Boudon, S. Topical use of dexpanthenol: A 70th anniversary article. *J. Dermatol. Treat.* **2017**, *28*, 766–773. [CrossRef]

20. Pavaloiu, R.D.; Stoica, A.; Stroescu, M.; Dobre, T. Controlled release of amoxicillin from bacterial cellulose membranes. *Cent. Eur. J. Chem.* **2014**, *12*, 962–967. [CrossRef]

21. Gerasimenko, D.V.; Avdienko, I.D.; Bannikova, G.E.; Zueva, O.Y.; Varlamov, V.P. Antibacterial effects of water-soluble low-molecular-weight chitosans on different microorganisms. *Appl. Biochem. Microbiol.* **2004**, *40*, 253–257. [CrossRef]

22. Zou, P.; Yang, X.; Wang, J.; Li, Y.; Yu, H.; Zhang, Y.; Liu, G. Advances in characterisation and biological activities of chitosan and chitosan oligosaccharides. *Food Chem.* **2016**, *190*, 1174–1181. [CrossRef] [PubMed]

23. Miller, L.S.; Cho, J.S. Immunity against *Staphylococcus aureus* cutaneous infections. *Nat. Rev. Immunol.* **2011**, *11*, 505–518. [CrossRef]

24. Simões, D.; Miguel, S.P.; Ribeiro, M.P.; Coutinho, P.; Mendonça, A.G.; Correia, I.J. Recent advances on antimicrobial wound dressing: A review. *Eur. J. Pharm. Biopharm.* **2018**, *127*, 130–141. [CrossRef] [PubMed]

25. Foster, E.J.; Moon, R.J.; Agarwal, U.P.; Bortner, M.J.; Bras, J.; Camarero-Espinosa, S.; Chan, K.J.; Clift, M.J.D.; Cranston, E.D.; Eichhorn, S.J.; et al. Current characterization methods for cellulose nanomaterials. *Chem. Soc. Rev.* **2018**, *47*, 2609–2679. [CrossRef] [PubMed]

26. Bellamy, L.J. *The Infrared Spectra of Complex Molecules*, 3rd ed.; Chapman and Hall, Ltd.: London, UK, 1975; ISBN 0412138506.

27. Tamizi, E.; Azizi, M.; Dorraji, M.S.S.; Dusti, Z.; Panahi-Azar, V. Stabilized core/shell PVA/SA nanofibers as an efficient drug delivery system for dexpanthenol. *Polym. Bull.* **2018**, *75*, 547–560. [CrossRef]

28. Fernandes, S.C.M.; Alonso-Varona, A.; Palomares, T.; Zubillaga, V.; Labidi, J.; Bulone, V. Exploiting mycosporines as natural molecular sunscreens for the fabrication of UV-absorbing green materials. *ACS Appl. Mater. Interfaces* **2015**, *7*, 16558–16564. [CrossRef]

29. Chiaoprakobkij, N.; Seetabhawang, S.; Sanchavanakit, N.; Phisalaphong, M. Fabrication and characterization of novel bacterial cellulose/alginate/gelatin biocomposite film. *J. Biomater. Sci. Polym. Ed.* **2019**, *30*, 961–982. [CrossRef]

30. Brown, J.E.; Davidowski, S.K.; Xu, D.; Cebe, P.; Onofrei, D.; Holland, G.P.; Kaplan, D.L. Thermal and structural properties of silk biomaterials plasticized by glycerol. *Biomacromolecules* **2016**, *17*, 3911–3921. [CrossRef]

31. Wu, C.-N.; Fuh, S.-C.; Lin, S.-P.; Lin, Y.-Y.; Chen, H.-Y.; Liu, J.-M.; Cheng, K.-C. TEMPO-oxidized bacterial cellulose pellicle with silver nanoparticles for wound dressing. *Biomacromolecules* **2018**, *19*, 544–554. [CrossRef]

32. Jana, S.; Trivedi, M.K.; Tallapragada, R.M.; Branton, A.; Trivedi, D.; Nayak, G.; Mishra, R.K. Characterization of physicochemical and thermal properties of chitosan and sodium alginate after biofield treatment. *Pharm. Anal. Acta* **2015**, *6*, 1000430. [CrossRef]

33. Silva, N.H.C.S.; Rodrigues, A.F.; Almeida, I.F.; Costa, P.C.; Rosado, C.; Neto, C.P.; Silvestre, A.J.D.; Freire, C.S.R. Bacterial cellulose membranes as transdermal delivery systems for diclofenac: In vitro dissolution and permeation studies. *Carbohydr. Polym.* **2014**, *106*, 264–269. [CrossRef]

34. Trovatti, E.; Silva, N.H.C.S.; Duarte, I.F.; Rosado, C.F.; Almeida, I.F.; Costa, P.; Freire, C.S.R.; Silvestre, A.J.D.; Neto, C.P. Biocellulose membranes as supports for dermal release of lidocaine. *Biomacromolecules* **2011**, *12*, 4162–4168. [CrossRef] [PubMed]

35. Tanrıverdi, S.T.; Suat, B.; Azizoğlu, E.; Köse, F.A.; Özer, Ö. In-vitro evaluation of dexpanthenol-loaded nanofiber mats for wound healing. *Trop. J. Pharm. Res.* **2018**, *17*, 387–394. [CrossRef]

36. Marques, M.R.C.; Loebenberg, R.; Almukainzi, M. Simulated biological fluids with possible application in dissolution testing. *Dissolut. Technol.* **2011**, *18*, 15–28. [CrossRef]

37. Costa, P.; Lobo, J.M.S. Modeling and comparison of dissolution profile. *Eur. J. Pharm. Sci.* **2001**, *13*, 123–133. [CrossRef]

38. Dubey, D.; Malviya, R.; Sharma, P.K. Mathematical modelling and release behaviour of drug. *Drug Deliv. Lett.* **2014**, *4*, 254–268. [CrossRef]

39. Maderuelo, C.; Zarzuelo, A.; Lanao, J.M. Critical factors in the release of drugs from sustained release hydrophilic matrices. *J. Control. Release* **2011**, *154*, 2–19. [CrossRef] [PubMed]

40. Feng, J.; Shi, Q.; Li, W.; Shu, X.; Chen, A.; Xie, X.; Huang, X. Antimicrobial activity of silver nanoparticles in situ growth on TEMPO-mediated oxidized bacterial cellulose. *Cellulose* **2014**, *21*, 4557–4567. [CrossRef]

41. Helaly, G.F.; El-Aziz, A.A.A.; Sonbol, F.I.; El-Banna, T.E.; Louise, N.L. Dexpanthenol and propolis extract in combination with local antibiotics for treatment of Staphylococcal and Pseudomonal wound infections. *Arch. Clin. Microbiol.* **2011**, *2*, 1–15. [CrossRef]

42. Wu, J.; Xiao, Z.; Chen, A.; He, H.; He, C.; Shuai, X.; Li, X.; Chen, S.; Zhang, Y.; Ren, B.; et al. Sulfated zwitterionic poly(sulfobetaine methacrylate) hydrogels promote complete skin regeneration. *Acta Biomater.* **2018**, *71*, 293–305. [CrossRef]

43. Saïdi, L.; Vilela, C.; Oliveira, H.; Silvestre, A.J.D.; Freire, C.S.R. Poly(N-methacryloyl glycine)/nanocellulose composites as pH-sensitive systems for controlled release of diclofenac. *Carbohydr. Polym.* **2017**, *169*, 357–365. [CrossRef]

44. Vilela, C.; Oliveira, H.; Almeida, A.; Silvestre, A.J.D.; Freire, C.S.R. Nanocellulose-based antifungal nanocomposites against the polymorphic fungus Candida albicans. *Carbohydr. Polym.* **2019**, *217*, 207–216. [CrossRef]

45. Chantereau, G.; Sharma, M.; Abednejad, A.; Vilela, C.; Costa, E.M.; Veiga, M.; Antunes, F.; Pintado, M.M.; Sèbe, G.; Coma, V.; et al. Bacterial nanocellulose membranes loaded with vitamin B-based ionic liquids for dermal care applications. *J. Mol. Liq.* **2020**, *302*, 112547. [CrossRef]

46. ISO 10993-5:2009(E). *Biological Evaluation of Medical Devices—Part. 5: Tests for In Vitro Cytotoxicity*; ISO: Geneva, Switzerland, 2009.

47. Xiang, J.; Shen, L.; Hong, Y. Status and future scope of hydrogels in wound healing: Synthesis, materials and evaluation. *Eur. Polym. J.* **2020**, *130*, 109609. [CrossRef]

48. Sharma, M.; Sahu, K.; Singh, S.P.; Jain, B. Wound healing activity of curcumin conjugated to hyaluronic acid: In vitro and in vivo evaluation. *Artif. Cells Nanomed. Biotechnol.* **2018**, *46*, 1009–1017. [CrossRef]

49. Johnston, S.T.; Ross, J.V.; Binder, B.J.; Sean McElwain, D.L.; Haridas, P.; Simpson, M.J. Quantifying the effect of experimental design choices for in vitro scratch assays. *J. Theor. Biol.* **2016**, *400*, 19–31. [CrossRef]

Publisher's Note: MDPI stays neutral with regard to jurisdictional claims in published maps and institutional affiliations.

nanomaterials

Article

The Effect of High Lignin Content on Oxidative Nanofibrillation of Wood Cell Wall

Simon Jonasson [1], Anne Bünder [2], Linn Berglund [1], Magnus Hertzberg [3], Totte Niittylä [2] and Kristiina Oksman [1,4,*]

1 Division of Materials Science, Department of Engineering Sciences and Mathematics, Luleå University of Technology, SE 97187 Luleå, Sweden; simon.jonasson@ltu.se (S.J.); linn.berglund@ltu.se (L.B.)
2 Umeå Plant Science Centre, Department of Forest Genetics and Plant Physiology, Swedish University of Agricultural Sciences, SE 90187 Umeå, Sweden; anne.bunder@slu.se (A.B.); totte.niittyla@slu.se (T.N.)
3 SweTree Technologies AB, SE 90403 Umeå, Sweden; magnus.hertzberg@swetree.com
4 Department of Mechanical & Industrial Engineering, University of Toronto, Toronto, ON M5S 3G8, Canada
* Correspondence: kristiina.oksman@ltu.se; Tel.: +46-920-497731

Abstract: Wood from field-grown poplars with different genotypes and varying lignin content (17.4 wt % to 30.0 wt %) were subjected to one-pot 2,2,6,6-Tetramethylpiperidin-1-yl)oxyl catalyzed oxidation and high-pressure homogenization in order to investigate nanofibrillation following simultaneous delignification and cellulose oxidation. When comparing low and high lignin wood it was found that the high lignin wood was more easily fibrillated as indicated by a higher nanofibril yield (68% and 45%) and suspension viscosity (27 and 15 mPa·s). The nanofibrils were monodisperse with diameter ranging between 1.2 and 2.0 nm as measured using atomic force microscopy. Slightly less cellulose oxidation (0.44 and 0.68 mmol·g^{-1}) together with a reduced process yield (36% and 44%) was also found which showed that the removal of a larger amount of lignin increased the efficiency of the homogenization step despite slightly reduced oxidation of the nanofibril surfaces. The surface area of oxidized high lignin wood was also higher than low lignin wood (114 m^2·g^{-1} and 76 m^2·g^{-1}) which implicates porosity as a factor that can influence cellulose nanofibril isolation from wood in a beneficial manner.

Keywords: cellulose nanofibrils; wood; lignin; TEMPO-oxidation

check for **updates**

Citation: Jonasson, S.; Bünder, A.; Berglund, L.; Hertzberg, M.; Niittylä, T.; Oksman, K. The Effect of High Lignin Content on Oxidative Nanofibrillation of Wood Cell Wall. *Nanomaterials* **2021**, *11*, 1179. https://doi.org/10.3390/nano11051179

Academic Editor: Carla Vilela

Received: 23 March 2021
Accepted: 21 April 2021
Published: 29 April 2021

Publisher's Note: MDPI stays neutral with regard to jurisdictional claims in published maps and institutional affiliations.

1. Introduction

The valorization of wood and other lignocelluloses generally include a variety of costly mechanical-chemical refining steps in order to overcome the inherent recalcitrance that comes from lignified cell walls and complex inaccessible crystalline cellulose microfibril architectures [1]. Effectively nanofibrillating the cell wall is an attractive prospect where cellulose nanofibrils (CNFs) are attainable as a final product or as an intermediate for further treatments that utilizes the increased accessibility [2]. For instance, through improved saccharification of nanofibrils compared to initial wood fibers [3]. An important factor that has made CNF isolation more favorable is the implementation of various pretreatments in addition to traditional pulping processes. These swell the cell wall structure and reduces its structural integrity, thereby making it more susceptible to complete nano-fibrillation. The choice of pretreatment, cellulosic feedstock and mechanical disintegration method varies across literature and is a subject of intense research activity [4]. A common process in this context is that of 2,2,6,6-Tetramethylpiperidin-1-yl)oxyl (TEMPO) catalyzed oxidation. This pretreatment mediates nanofibrillation through selective carboxylation of the cellulose C$_6$-hydroxyl groups through N-oxyl radical generation [5]. The treatment has been studied in great depth in relation to delignified pulps from a variety of sources [6]. It has also been demonstrated that feedstock containing lignin, such as thermomechanical pulp [7], hemp [8] and wood [9], can be simultaneously delignified and carboxylated using the treatment. The relation between these procedures and nanofibrillation is complex where CNFs with residual lignin are attainable despite substantial carboxylation [10,11]. It has also been reported that the presence of the TEMPO-catalyst significantly increases carboxylation whilst keeping the delignifying efficiency constant [7]. The understanding of how these phenomena occurs in relation to native wood structures with varying lignin is currently unexplored. Available literature revolves around already processed feedstocks or focuses on pure mechanical fibrillation [12]. A natural step would thus be to combine TEMPO-oxidation with variation in lignin content of unprocessed feedstocks. The associated novelty includes highlighting TEMPO-oxidation as a metric of fibrillation suitability for raw wood with variation in lignin content, something which could be applicable to commercial endeavors where TEMPO-oxidation is an attractive prospect in obtaining CNFs. Direct TEMPO-oxidation of wood has only recently been the subject of few dedicated studies [9,13]. These studies were furthermore showcasing one type of wood whereas the study presented herein aims to explore an additional factor of wood characteristics. This can increase the knowledge and feasibility of direct oxidation of wood by elucidating how wood properties, in this case native lignin content, influences the process.

In this study, we investigated field-grown poplar trees with large (17.4–30.0 wt %) systematic differences in lignin content. The goal was to analyze how lignin content associated wood recalcitrance is manifested in relation to nanofibrillation after pretreatments that simultaneously delignifies and carboxylates the wood fibers. Increased understanding of how oxidative treatments interplay with variation in lignin content of the feedstock is beneficial to a wide range of CNF isolation endeavors through elucidation of the link between wood and final CNFs.

2. Materials and Methods

2.1. Materials

Six different wood samples with specific chemical composition (see Table 1) were supplied by SweTree Technologies AB, Umeå Sweden. The samples were selected out of a library of trees (see Supplementary Material) based on the largest variety of total lignin content as determined using pyrolysis GS/MS [14]. The variation was validated using principal component analysis (see Supplementary Material). Three wood samples with similar lignin content were chosen as center points, approximately corresponding to lignin content in between the lowest and the highest samples. This was done in order to be able to evaluate trends between lignin content and characterization if present.

Table 1. Composition of the wood feedstock used in this study shown in wt %.

Sample ID	Coding	Carbohydrates	Extractives	Total Lignin	Cellulose [1]
1	Low	79.4	0.4	17.4	45 (4)
2		77.2	0.4	19.7	42 (3)
3	Medium	72.4	0.7	24.4	41 (2)
4		72.3	0.5	24.5	40 (3)
5		72.5	0.6	24.6	41 (2)
6	High	66.9	0.7	30.0	39 (2)

[1] as determined using Test Methods T 203 C TAPPI-standard [15] with three technical replicates on the basis of dried chlorite holocellulose.

High-purity sodium chlorite (77.5–82.5%), standard hydrochloric acid solution (0.5 N), standard sodium hydroxide solution (0.1 N), sodium hydroxide beads ($\geq$97%, ACS) were purchased from VWR, Solna, Sweden. TEMPO (99%), sodium hypochlorite (NaClO, 6–14% active chlorine) and Congo red (analytical standard) were purchased from Sigma-Aldrich, Stockholm, Sweden AB. All chemicals were used as received.

2.2. Oxidation and Homogenisation

Wood powder of the grinded poplar trees were sieved through a 300–500 μm mesh prior to sampling (2 g) and soaking in (100 mL) distilled water for 24 h prior to further treatment. Primary oxidant sodium chlorite 5.0 $g \cdot g^{-1}$ wood and TEMPO (17.5 $mg \cdot g^{-1}$ wood) was added together with a phosphate buffer (0.1 M, 100 mL, pH = 6.8) to the vessel (250 mL) containing the wet powder. Each vessel with respective suspension was closed and submerged into a SBS40 shaking water bath (Cole-Parmer Stuart, Staffordshire, United Kingdom) at 60 °C for 1 h at 200 rpm in order to dissolve $NaClO_2$ and TEMPO. Shaking water bath was used to eliminate experimental error from variation in heating rate and stirring and to make sure each tree sample was subjected to identical processing conditions. The reaction was initiated through addition of sodium hypochlorite (2 $mL \cdot g^{-1}$ wood) and then left for 48 h. The methodology was adapted from [16] with changes to the amount of primary oxidant and feedstock characteristics. The fully delignified and carboxylated wood was filtered thoroughly and then homogenized one pass without recirculation using an APV-2000 high-pressure homogenizer (SPX Flow Inc, Silkesborg, Denmark) with an average flow rate of 4 $mL \cdot s^{-1}$ at a pressure of 1000 bar. The concentration of respective sample was adjusted prior to and after homogenization to eliminate unwanted effects from variation in solid content.

2.3. Viscosity of CNF Suspensions

The viscosity of the oxidized and homogenized suspensions was measured at the same concentrations (0.183 ± 0.01 wt %) using a Vibro Viscometer (SV-10, A&D Company Limited, Tokyo, Japan) with a tuning fork vibration method at a vibrational frequency of 30 Hz. The measurements were repeated three times at room temperature for each sample.

2.4. Carboxylate Content

Carboxylate content was analyzed through the electric conductivity titration method adapted from [17]. 150 mL of respective CNF (cellulose nanofibril) suspensions ($\approx$0.2 wt %) were protonated through addition of 0.1 M hydrochloric acid and 0.01 M sodium chloride. The suspensions were titrated with fresh 0.01 M sodium hydroxide through 0.5 mL increments until pH 10 was reached. The amount of carboxylate groups ($mmol \cdot g^{-1}$) induced from TEMPO-oxidation was calculated through Equation (1) below where c is the concentration of the sodium hydroxide, V_2 and V_1 is the volume of added sodium hydroxide at the end and start, respectively, m is the mass of the cellulosic material in the sample, calculated through subtraction of added acid, base and salt mass from the oven dried suspension. The titration was repeated three times.

$$\text{Carboxylate groups} \left(\frac{\text{mmol}}{\text{g}} \right) = \frac{C(V_2 - V_1)}{m} \tag{1}$$

2.5. Atomic Force Microscopy

Tapping-mode atomic force microscopy (AFM) Veeco MultiMode scanning probe, Santa Barbara, CA, USA, was used to confirm the presence of nanofibrils and analyze the morphology. Antimony-doped silicon cantilevers (TESPA-V2, Bruker, Camarillo, CA, USA) with a spring constant of 42 $N \cdot m^{-1}$ and a nominal tip radius of 8 nm were used for the analysis. Samples were prepared by depositing a small droplet of the CNF suspension (0.001 wt %) on a freshly cleaved mica plate and letting it air dry for $\geq$ 5 h. Around 100 fully individualized nanofibrils, total from four micrographs were analyzed for each sample and presented as a mean.

2.6. Process and Nanofibril Yield

Process yield was calculated as the washed gravimetric yield after the chemical treatment relative dry wood mass. The fraction of the process yield that comprised individual, colloidally stable nanofibrils were further quantified through centrifugation of the suspensions at 12,000$\times$ g (Avanti J25i, Beckman Coulter Inc. Brea, CA, USA) for 20 min at an approximate consistency of 0.2 wt %. The suspensions were decanted and the solids retained in the sediment were dried for 24 h at 95 °C. This was repeated three times and the nanofibril-fraction was then calculated according to the Equation (2), where m_p and m_s is the dry precipitate and supernatant mass in the sample, respectively. The nanofibril-fraction (Φ) was presented as fraction of the process yield.

$$\Phi = 1 - \frac{m_p}{m_p + m_s} \tag{2}$$

2.7. Thermogravimetric Analysis

Dried CNF suspensions were subjected to thermogravimetric analysis TGA-Q500, TA Instruments, (New Castle, DE, USA) to analyze the relative amount of residue after pyrolysis of the oxidized wood samples and to compare the thermal stability of the CNFs. A heating rate of 10 °C·min^{-1} from room temperature to 900 °C was used in a nitrogen atmosphere.

2.8. Klason Lignin and Cellulose Content

Cellulose content of the wood holocellulose and dried CNF suspensions were calculated based on extraction using 17.5 M NaOH according to TAPPI-standard [15], where the oven dried gravimetric yield after filtration was used to estimate α-cellulose content. Klason lignin for the dry CNF suspensions was analyzed according to standard [18] through hydrolysis using sulfuric acid

2.9. Network Manufacturing

CNF suspensions were degassed for 30 min in a vacuum oven prior to vacuum filtration on hardened filter paper (Whatman, Grade 52, GE Healthcare, Machelen, Belgium, Pore size: 7 μm). The wet networks were carefully peeled from the filter paper and dried to around 14 % solid content prior. The stable networks were placed between two blotting papers and pressed (2 kN·m^{-2}) for 10 h. The dry networks were further compression molded using Fontijne Grotnes LPC-300 (Vlaardingen, Netherlands), between two mylar films (Lohmann Technologies, Knowl Hill, UK) at a pressure of 0.32 MPa and at a temperature of 120 °C.

2.10. X-ray Diffraction

An X-ray diffractometer (PANalytical, Almelo, Netherlands) equipped with a PIXcel^{3d} detector was used to measure the degree of crystallinity of the dried nanofibrils. Cu-Kα radiation (λ = 0.154 nm) was used during analysis with operation voltage of 45 kV and a current of 40 mA. The crystallinity index (CI) was calculated according to the peak height method after baseline correction [19]. This was done according to Equation (3), where I_{200} is the intensity of the crystalline (200) peak at 2θ = 22.4 and I_{am} is the intensity minimum between nearby crystalline peaks at 2θ = 18.

$$CI \, (\%) = \frac{I_{200} - I_{am}}{I_{200}} \times 100 \tag{3}$$

2.11. Mechanical Testing

The dry networks were cut into rectangular samples (40 mm × 5 mm) using a mechanical punch. The samples were then stored at 50% RH for at least two days prior to mechanical characterization. Mechanical testing was performed using a Shimadzu AG-X universal testing machine (Kyoto, Japan) with a 500 N load cell. Testing was performed at a crosshead speed of 10%·min^{-1}, with the strain being measured using a video extensometer (high-speed camera, HPV-X2). The gauge length was set to 20 mm for each measurement. Seven specimens were analyzed for each network batch. The tensile strength was reported as the maximum strength at break. Young's modulus was calculated from the slope of the stress–strain curve at early strain (0.1–0.5 %).

2.12. Sample Porosity and Moisture Content

Density of the networks were measured by calculation based on the specimen volume and weight. Porosity (P) was derived using Equation (4), where ρ_s and ρ_c is the density of the specimen and cellulose (1.5 g·cm^{-3}), respectively [20] Moisture content of the specimen was calculated from difference in weight after 24 h in oven at 105 °C

$$P = 1 - \frac{\rho_s}{\rho_c} \tag{4}$$

2.13. Wood Porosity Analysis

Intact wood samples at the extreme values of initial lignin content (17.4 and 30.0%) were oxidized in an identical manner as described in Section 2.2. The oxidized samples were carefully washed by submerging the samples into 20 L of distilled water. The fragile delignified and carboxylated samples were left to soak for four hours and the process was repeated three times with fresh distilled water. Brunauer–Emmett–Teller (BET) specific surface area analysis of the samples were performed by first freezing the wet samples at −20 °C for 10 h prior to freeze-drying for 72 h using Alpha 1-2 LD plus (Martin Christ Gefriertrocknungsanlagen GmbH, Osterode am Harz, Germany). The samples were degassed overnight at 120 °C followed by N_2 physiosorption using a Micromeritics Gemini VII surface area analyzer (Norcross, GA, USA). A total of 30–60 mg of oxidized wood sample was analyzed for a respective sample.

Surface area of the oxidized wood was also estimated in the wet state by analyzing congo red absorption of the cellulosic fibers as described by [12]. Briefly, the absorbed dye was quantified using a UV-vis spectrophotometer (GENESYS, 10 UV, Thermo Scientific, Schwerte, Germany) at the absorption maxima 500 nm and then translated to surface area (SSA) using Equations (5) and (6), where E is the solution concentration of congo red at equilibrium (mg·mL^{-1}), A_{max} is the absorbed amount of congo red on the sample (mg·g^{-1}), K_{ad} is the equilibrium constant, N is Avogadro's constant, SA is the theoretical surface area of a congo red molecule (1.73 nm^2) and MW is the molecular weight of congo red (697 g·mole^{-1}).

$$\frac{E}{A} = \frac{1}{K_{ad}\,A_{max}} + \frac{E}{A_{max}} \tag{5}$$

$$SSA = \frac{A_{max} \times N \times SA}{MW \times 10^{21}} \tag{6}$$

Samples cut in the wet state were freeze-dried and analyzed using a scanning electron microscope JEOL (JSM-IT300, Tokyo, Japan) at an acceleration voltage of 5 kV. A 15 nm layer of platinum was sputtered on the samples prior to analysis.

3. Results and Discussion

3.1. Fibrillation Efficiency and Nanofibril Characteristics

Suspensions of oxidized and homogenized wood samples showed a difference in viscosity between 9 and 27 mPa·s (shown in Figure 1a) is indicative of a qualitatively higher degree of fibrillation [21]. A similar behavior is reflected in the fraction of the yield that comprises fine colloidally stable nanofibrils, (shown in Figure 1b). A total of 45–68 wt % of the yield was recovered upon fractionating. A higher recovery of nanofibrils from wood with higher lignin content can be seen which is supported by the increased viscosity. The process yield ranged between 36 and 44 wt % of dry initial wood and decreased with more initial lignin content of the wood.

This can be understood from the complete dissolution of the lignin after being subjected to the catalytic conditions in this study. Klason lignin difference was also negligible after the treatment where all samples had an estimated residual lignin content of <1 wt %, indicating that a full dissolution of lignin took place following oxidation. A similar enhanced delignification has been observed after treatment of thermomechanical pulp using the related TEMPO/NaBr/NaClO-system [7]. It is further noteworthy that the nanofibril yield relative initial dry wood is similar for high lignin containing wood despite lower relative cellulose content. This is indicative of effects associated with higher initial lignin content and is the subject of discussion in Section 3.3.

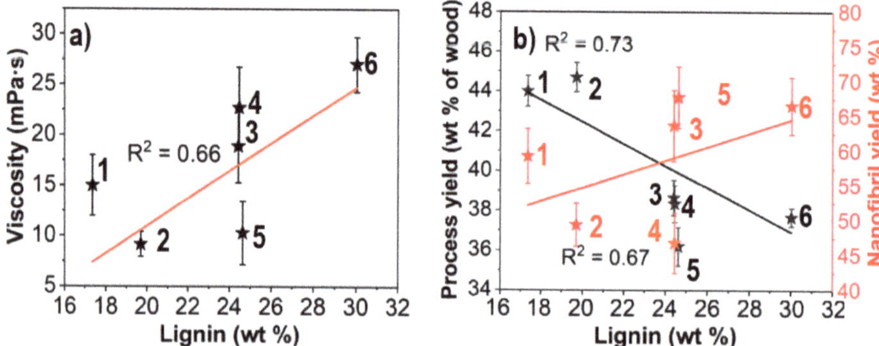

Figure 1. Viscosity (**a**), process yield (**b**) and nanofibril yield (**b**, red) of CNF suspensions obtained from wood with varying initial lignin content. Data represent mean ± SD of six different CNF samples (*n* = 3 replicates per sample). Data points for respective samples indicated with black and red stars.

Carboxylate content of the final suspensions is shown in Figure 2 together with AFM micrographs of the CNFs that comprises the suspensions. Carboxylate content ranged between 0.44 and 0.68 mmol·g^{-1} of cellulose and was higher for low lignin wood compared to high. The range of the height of the individualized CNFs was measured between 1.2 and 2.0 nm and are thus thinner than most models suggest for elementary fibrils of wood cellulose. Significant differences in height were only apparent when comparing low lignin wood (sample #1) to the rest.

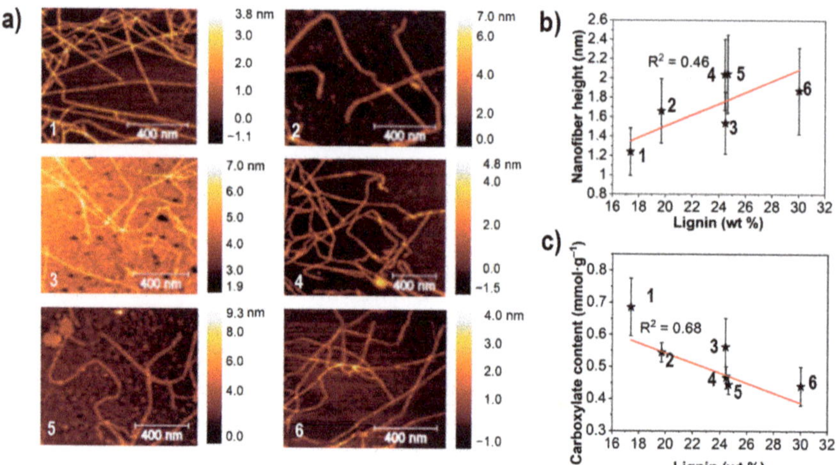

Figure 2. AFM micrographs (**a**) of the CNFs with corresponding height (**b**) and carboxylate content (**c**). Error bars in (**b**) derived from standard deviation of the size distribution. Data represent mean ± SD of six different CNF samples with 50–100 measurements per sample. Data points for respective samples indicated with black stars.

The relation between carboxylate content and initial lignin content indicates a decrease in carboxylation with more initial lignin of the feedstock although the correlation cannot be extrapolated to a larger range of carboxylation due to a rather narrow interval that the carboxylation ranges within. Interestingly, at 0.37 mmol·g^{-1} it has been calculated that all the surfaces of native cellulose microfibrils have undergone oxidation (Sjöstedt, 2014), which implies that the smaller individual elementary fibrils comprising these microfibrils are the subjects of a variable oxidation throughout this study. This can be explained by looking at the oxidation process where lignin hydroxyl groups are efficiently oxidized and depolymerized by TEMPO-systems through cleaving of C-C and ether bonds [22]. Lignin is present in both the thin middle lamella and the cell wall layers of wood [23] and can expectedly be preferentially oxidized during the process. This would decrease the effectiveness of the cellulose oxidation since lignin is more accessible than cellulose and the process is defined by a specified amount of chlorite (5 g·g^{-1} wood). This has been reported previously [24] and explained through splitting of the elementary fibril into smaller nanofibrils and even individual polymer chains [25]. This fragmentation phenomenon has been reported by multiple authors [24–27]. The increased carboxylation taking place for the cellulose derived from low lignin wood would according to this hypothesis be related to the decreased height of the CNFs from low lignin wood as shown in Figure 2. The correlation between the size distribution as obtained from AFM and other metrics of fibrillation is furthermore more accurate for highly fibrillated samples as in this study [28].

From the AFM analysis there were also particles present across the samples, especially apparent in CNFs made from high lignin wood. A larger scale AFM micrograph of that sample is shown in Figure 3, where particles and clusters are scattered among the CNFs.

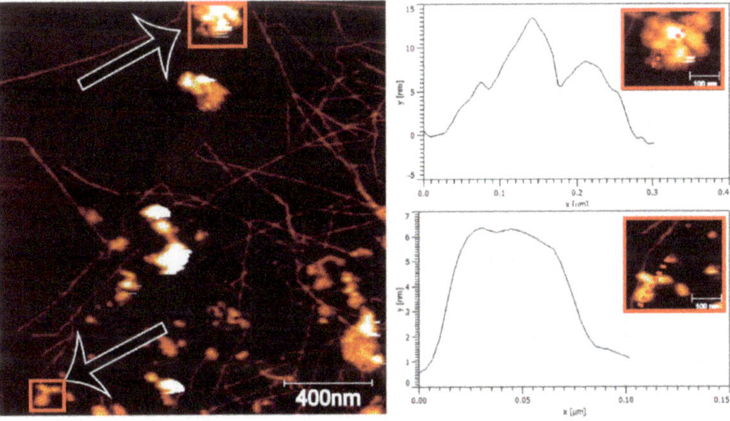

Figure 3. AFM micrograph of CNF (**left**) made from high lignin wood with particles and clusters of irregular shapes highlighted (**right**).

This is hypothesized to be residues from processing. Even though extensive filtration took place during the isolation process it is likely that trapping of non-cellulosic components occurs inside the cell wall during oxidation and liberated together with the CNFs during homogenization. This is further supported by an increase in char after pyrolysis of the samples as shown in Figure 4. A mass fraction between 0.16 and 0.38 was remaining after pyrolysis of nanofibrils with more residues remaining for CNFs made from high lignin wood. The thermal stability of all CNFs was also similar as shown in Figure 4 and this is in agreement with what commonly is reported for CNFs derived from the TEMPO/NaClO/NaClO$_2$-treatment.

Pyrolysis of lignin generally leads to a higher char formation compared to cellulose and hemicellulose [29]. In the absence of lignin, however, as shown for these samples, the residues have to originate from another source. Higher amount of char formation from the pyrolysis of deprotonated CNFs have been reported, where a difference up to 16 wt % was reported solely from the deprotonation of the carboxylate groups [30]. This effect together with the presence of residues from the oxidation explains this phenomenon and is further elaborated upon in Section 3.3 with porosity analysis of the TEMPO-oxidized wood. Thermal degradation occurs earlier compared to chlorite bleached wood, but later compared to cellulose that has undergone the harsher TEMPO/NaBr/NaClO-treatment [27]. The lack of variation amongst the samples is also supported by the carboxylate content (Figure 2), where the differences are relatively small in comparison to the whole range of carboxylate content that is possible to obtain after TEMPO-oxidation of all available hydroxyl groups on the nanofibril surfaces.

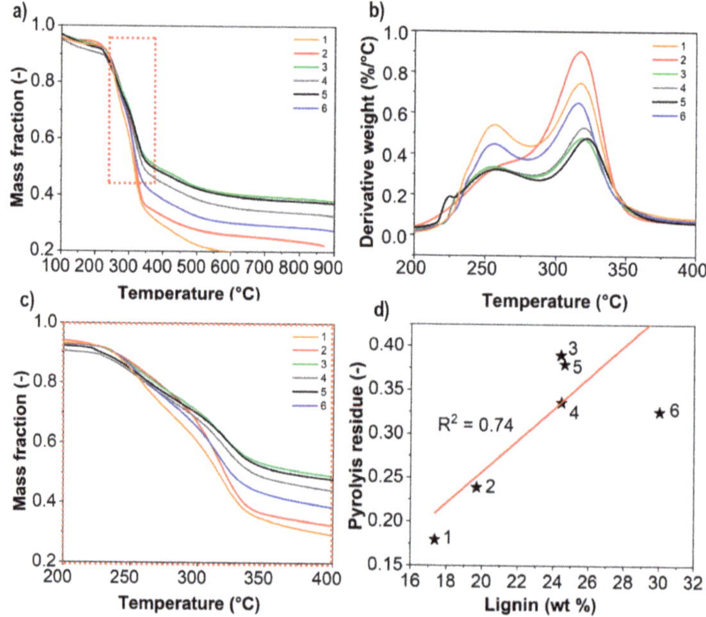

Figure 4. Thermal degradation curves showing mass loss (**a**), derivative weight (**b**) mass loss curve with zoom-in (200–400 °C) (**c**) and corresponding mass fraction of residues remaining at 900 °C (**d**) Data points for respective samples indicated with black stars.

3.2. Dry Network Characteristics

X-ray diffractograms of dried networks and corresponding degree of crystallinity is presented in Figure 5. The degree of crystallinity of the networks was similar at a range between 77 and 82% with a slight decrease from low to high lignin wood. The native cellulose I_β crystal structure was retained after processing for all samples, as indicated by the diffractogram peak location [31]. Both the absolute value of crystallinity and tendencies for a decrease in crystallinity as a function of fibrillation agree with previous analysis of a wide range of CNFs [32].

The presence of moieties responsible for variation in char formation (as shown in Figure 4) seems to have no effect on crystallinity. This is in accordance with the general behavior of TEMPO-oxidized celluloses [17] and is supported by reports that indicate that degree of crystallinity is not related to the relative amount of cellulose in a given sample [30].

The mechanical properties of the networks are presented in Figure 6 and shows specific strength that is constant with respect to lignin content. A slight increase can be seen for the middle samples, but not indicative of any trend. Specific modulus increased with initial lignin content between 3 and 6 GPa, whereas the elongation-at-break decreased from 8.2% to 4.2%. The opposite trends for modulus and elongation-at-break are noteworthy and is indicative of the presence of previously mentioned impurities. The presence of stiff particles would significantly increase the stiffness of the networks whilst at the same time decreasing the elongation-at-break by acting as defects and limiting stress transfer between the CNFs [33].

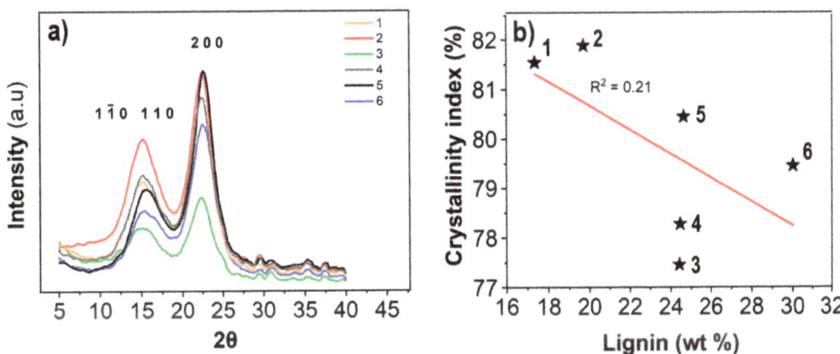

Figure 5. X-ray diffractograms of dry CNF networks (**a**) with corresponding Segal crystallinity indices (**b**). Data points for respective sample are indicated with black stars.

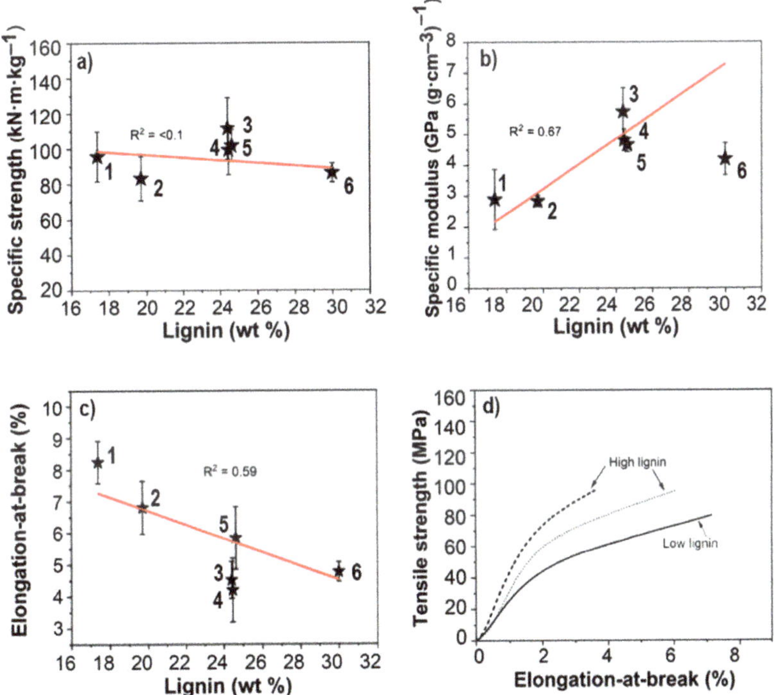

Figure 6. Mechanical properties of dry CNF networks with specific strength (**a**), specific modulus (**b**) and elongation-at-break (**c**). Representative stress–strain curves that showcase the difference between networks made from low and high lignin wood (**d**). Data represents mean ± SD of six different network samples (*n* = 7 replicates per sample). Data points for respective sample are indicated with black stars.

Additional characteristics that are measurable and may influence the behavior of the dry networks are porosity/density, moisture content and relative amount of cellulose [20] This data is presented in Table 2 and shows that there are no obvious differences in physical characteristics of the specimen that may have been responsible for skewing the mechanical properties of the dry networks. The presence of a relatively large fraction of hemicellulose (24–31 wt %) is also interesting and is in agreement with our previous reports on similar processing for poplar wood [27] where an α-cellulose content of 70 wt % was obtained after direct TEMPO/NaClO/NaClO$_2$ treatment.

Table 2. Physical properties of the dry networks made from CNF suspensions. Data represents mean $\pm$ SD of six different network samples (*n* = 3 replicates per sample).

Sample ID	Moisture (wt %)	Density (g·cm^{-3})	Thickness (μm)	Porosity (%)	α-cellulose (wt %)	Lignin (wt %)
1	9.1 (1.1)	1.20 (0.10)	71 (10)	20.0 (6.7)	74	<1
2	7.9 (0.8)	1.23 (0.07)	63 (4)	18.0 (4.7)	71	<1
3	7.8 (1.3)	1.19 (0.09)	88 (7)	20.7 (6.0)	76	<1
4	6.8 (1.3)	1.26 (0.10)	50 (5)	16.0 (6.7)	70	<1
5	8.0 (1.4)	1.17 (0.11)	65 (6)	22.0 (7.3)	69	<1
6	7.5 (0.5)	1.20 (0.09)	78 (10)	20.0 (6.0)	76	<1

3.3. Wood Porosity Hypothesis

The results shown in this work have been argued in part on the basis of a higher porosity in the wood cell wall after oxidation of samples with a larger amount of initial lignin. Entrapment of subsequent treatment chemicals within the porous cell wall have also been connected to the increased amount of char following pyrolysis despite negligible amount of residual lignin. In order to substantiate this hypothesis further, intact pieces of wood were subjected to similar oxidation processes and investigated in terms of surface area following TEMPO-oxidation and electron microscopy of the wood surfaces. The surface area as estimated from BET and congo red adsorption is presented in Table 3 and indicates a larger surface area in the wet state (congo red) for high lignin wood (114 and 76 m^{2}g^{-1}) compared to low. The difference between both samples is negligible after freeze-drying and is lower compared to the wet state (4.4 and 6.9 m^{2}g^{-1}).

Table 3. Surface area, pore volume and pore size of oxidized wood samples (*n* = 1) as estimated in the freeze-dried state (BET) and in a never-dried state (Congo red).

Sample ID	Coding	BET Surface Area (m^{2}g^{-1})	Pore Volume (cm^3·kg^{-1})	Pore Size (nm)	Wet Surface Area (m^{2}g^{-1})
1	Low	6.9	7.5	4.4	76
6	High	4.4	5.7	5.0	114

The large difference in surface area estimated from BET and congo red adsorption can be explained by significant hornification occurring during freeze-drying and subsequent blocking of pores across different length scales whereas dye adsorption is performed for never-dried samples. Similar, relatively low values (14–42 m^{2}g^{-1}) have been reported for BET analysis of CNF aerogels [34], which was explained through irreversible agglomeration of the cellulosic surfaces during removal of water (hornification). A higher estimated surface area for dye adsorption compared to BET has also been reported for CNCs [35]. Hornfication for CNFs is, furthermore, more pronounced compared to CNCs [36], which further supports the increased differences observed in this study.

SEM micrographs of oxidized wood sections are shown in Figure 7 and further supports the observation of a higher surface area where oxidation of high lignin samples (Figure 7b) resulted in a fragile structure that appears to have been more affected by the oxidation (μm-mm scale) in both the longitudinal direction (Figure 7, upper) and the cross section (Figure 7, lower).

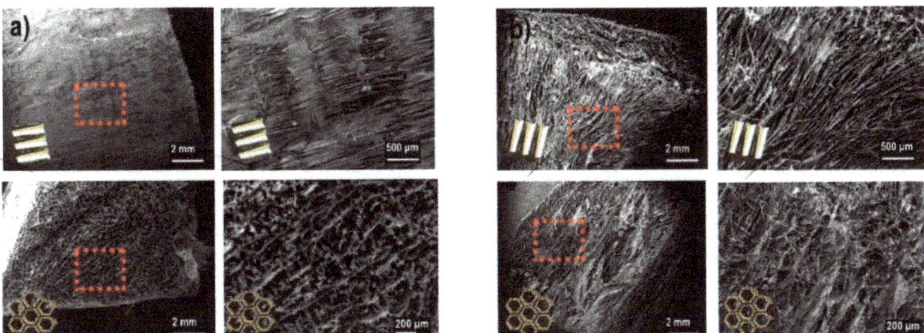

Figure 7. Structure of TEMPO-oxidized wood samples with low lignin wood (**a**) and high lignin wood (**b**). Upper figures showing the structure in the direction of the wood fibers and the lower showing corresponding cross sections.

This supports the observation of a more easily fibrillated high lignin wood after oxidation. This was also indicated when handling the wet delignified and oxidized wood samples where high lignin wood was more prone to breaking into smaller pieces. It should further be noted that the porosity induced from removal of lignin from the secondary cell wall was not directly visible during this analysis. Apart from the surface area estimations and large (μm-mm) scale porosity effects that have been shown in this study, direct imaging of nanoscale porosity following TEMPO-oxidation is likely beyond the resolution limit of this analysis.

3.4. Outlook and Limitations

In this work, linear regressions between lignin content and presented characteristics have been shown. In many cases a R^2-value of about 0.7 was apparent, which generally indicates good correlation to initial lignin content. Interesting data points to look at from the context of correlation are the three center points with three samples with approximately 24% initial lignin content. Apparent from the graphs shown throughout is that these tend to fluctuate and interchange in order of best performer. As such, it becomes difficult to correlate strictly between absolute value of lignin content of ease of fibrillation, more than looking at the extreme values of highest and lowest lignin content. The difference between the extremes is more certain and thus the subject of conclusion from this work. The reason behind uncertainty in absolute lignin content and fibrillation likely originate from the biological nature of wood where lignin content can be considered an observable characteristic that originates from the genetic background of the trees that are different across poplars used in this study. Another important emphasis is the choice of treatment that has been used. In this study, direct TEMPO-oxidation was used since it gives the opportunity to minimize the number of experimental steps (and errors) needed to go from wood to final CNFs. With the oxidative nature of the treatment the results are preferably generalized to similar treatments that both removes the lignin and modifies cellulose of the wood.

4. Conclusions

Native lignin content in poplar trees (17–30 wt %) was found to impact nanofibrillation and network characteristics following direct TEMPO-oxidation in neutral conditions. Process yields after the treatment ranged between 36 and 44 wt % of initial wood comprising of 45 to 68% nanofibrils with a corresponding viscosity increase between 9 and 27 mPa·s. These fibrillation metrics were higher for high lignin wood compared to low lignin wood whereas surface carboxylation was lower (0.44 mmol·g^{-1} and 0.68 mmol·g^{-1}). The height of the nanofibrils comprising the suspensions ranged between 1.2 and 2.0 nm as measured using AFM.

The dry networks made from the suspensions showed an increased stiffness (3–5 GPa), decreased elongation-at-break (8.2–4.2%) and a similar tensile strength (≈120 MPa) when made from high lignin wood. A higher fraction of char (0.16 to 0.38) was also produced after pyrolysis despite a similar, negligible amount of residual lignin (<1%).

The surface area of TEMPO-oxidized high lignin (30.0 wt %) wood was comparable in the hornified dry state (4.4 and 6.9 $m^2 \cdot g^{-1}$, BET) and higher in wet state (114 vs. 74 $m^2 \cdot g^{-1}$) compared to the low lignin (17.4 wt %).

Through the direct TEMPO oxidation of wood with a high lignin content, the delignification and carboxylation selectivity of the lignin leads to high porosity which is shown by the increased specific surface area and strongly affected wood ultrastructure. At very high lignin content, this approach effectively makes the wood cell wall more susceptible to disintegration, which in turn can be utilized to achieve a higher degree of fibrillation.

These findings highlight wood with high lignin content as being attractive for CNF isolation and challenges the general viewpoint that lignin is an impediment for processing of wood into CNFs. This study opens for further academic and commercial interest, where i) direct oxidation of wood is novel in the sense of quantifying wood suitability for nanofibrillation and ii) large scale processing can be investigated to look at the potential benefits of a similar lignin variation in a more upscaled commercial environment.

Supplementary Materials: The following are available online at https://www.mdpi.com/article/10.3390/nano11051179/s1, supplementary.docx: Figure S1: PCA of trees available in the library. Table S1: Composition of the trees as obtained from pyrolysis. Table S2: Library ID of trees studied in the manuscript.

Author Contributions: Conceptualization, S.J., L.B., K.O.; methodology, S.J.; validation, S.J., L.B. and K.O.; formal analysis, S.J.; investigation, S.J.; resources, K.O., T.N. and M.H.; data curation, S.J., A.B. and T.N.; writing—original draft preparation, S.J.; writing—review and editing, S.J., K.O., L.B. and T.N.; visualization, S.J.; supervision, K.O., L.B.; project administration, T.N. and K.O.; funding acquisition, T.N. and K.O. All authors have read and agreed to the published version of the manuscript.

Funding: Provided by FORMAS to the Nanowood project (942-2016-10) and Bio4Energy a Swedish Strategic Program.

Data Availability Statement: Data presented in this study are available on request from the corresponding author.

Acknowledgments: We are grateful for the financial support provided by FORMAS to the Nanowood project (942-2016-10) and for the supply and analysis of the poplar wood from SweTree Technologies, Umeå Sweden. Bio4Energy, The Swedish strategic research program and Kempestiftelserna are further acknowledged for support of infrastucture.

Conflicts of Interest: The authors declare no conflict of interest. The funders had no role in the design of the study; in the collection, analyses or interpretation of data; in the writing of the manuscript; or in the decision to publish the results.

References

1. Himmel, M.E.; Ding, S.-Y.; Johnson, D.K.; Adney, W.S.; Nimlos, M.R.; Brady, J.W.; Foust, T.D. Biomass recalcitrance: Engineering plants and enzymes for biofuels production. *Science* **2007**, *315*, 804–807. [CrossRef] [PubMed]
2. Lee, S.H.; Chang, F.; Inoue, S.; Endo, T. Increase in enzyme accessibility by generation of nanospace in cell wall supramolecular structure. *Bioresour. Technol.* **2010**, *101*, 7218–7223. [CrossRef]
3. Morales, L.O.; Iakovlev, M.; Martin-Sampedro, R.; Rahikainen, J.L.; Laine, J.; van Heiningen, A.; Rojas, O.J. Effects of residual lignin and heteropolysaccharides on the bioconversion of softwood lignocellulose nanofibrils obtained by SO2-ethanol-water fractionation. *Bioresour. Technol.* **2014**, *161*, 55–62. [CrossRef] [PubMed]
4. Nechyporchuk, O.; Belgacem, M.N.; Bras, J. Production of cellulose nanofibrils: A review of recent advances. *Ind. Crops Prod.* **2016**, *93*, 2–25. [CrossRef]
5. Shibata, I.; Isogai, A. Depolymerization of cellouronic acid during TEMPO-mediated oxidation. *Cellulose* **2003**, *10*, 151–158. [CrossRef]
6. Isogai, A.; Hänninen, T.; Fujisawa, S.; Saito, T. Review: Catalytic oxidation of cellulose with nitroxyl radicals under aqueous conditions. *Prog. Polym. Sci.* **2018**, *86*, 122–148. [CrossRef]

7.   Ma, P.; Fu, S.; Zhai, H.; Law, K.; Daneault, C. Influence of TEMPO-mediated oxidation on the lignin of thermomechanical pulp. *Bioresour. Technol.* **2012**, *118*, 607–610. [CrossRef]

8.   Puangsin, B.; Soeta, H.; Saito, T.; Isogai, A. Characterization of cellulose nanofibrils prepared by direct TEMPO-mediated oxidation of hemp bast. *Cellulose* **2017**, *24*, 3767–3775. [CrossRef]

9.   Jonasson, S.; Bünder, A.; Niittylä, T.; Oksman, K. Isolation and characterization of cellulose nanofibers from aspen wood using derivatizing and non-derivatizing pretreatments. *Cellulose* **2020**, *27*, 185–203. [CrossRef]

10.   Herrera, M.; Thitiwutthisakul, K.; Yang, X.; Rujitanaroj, P.O.; Rojas, R.; Berglund, L. Preparation and evaluation of high-lignin content cellulose nanofibrils from eucalyptus pulp. *Cellulose* **2018**, *25*, 3121–3133. [CrossRef]

11.   Wen, Y.; Yuan, Z.; Liu, X.; Qu, J.; Yang, S.; Wang, A.; Wang, C.; Wei, B.; Xu, J.; Ni, Y. Preparation and Characterization of Lignin-Containing Cellulose Nanofibril from Poplar High-Yield Pulp via TEMPO-Mediated Oxidation and Homogenization. *ACS Sustain. Chem. Eng.* **2019**, *7*, 6131–6139. [CrossRef]

12.   Spence, K.L.; Venditti, R.A.; Habibi, Y.; Rojas, O.J.; Pawlak, J.J. The effect of chemical composition on microfibrillar cellulose films from wood pulps: Mechanical processing and physical properties. *Bioresour. Technol.* **2010**, *101*, 5961–5968. [CrossRef]

13.   Kaffashsaie, E.; Yousefi, H.; Nishino, T.; Matsumoto, T.; Mashkour, M.; Madhoushi, M.; Kawaguchi, H. Direct conversion of raw wood to TEMPO-oxidized cellulose nanofibers. *Carbohydr. Polym.* **2021**, *262*, 117938. [CrossRef] [PubMed]

14.   Gerber, L.; Eliasson, M.; Trygg, J.; Moritz, T.; Sundberg, B. Multivariate curve resolution provides a high-throughput data processing pipeline for pyrolysis-gas chromatography/mass spectrometry. *J. Anal. Appl. Pyrolysis* **2012**, *95*, 95–100. [CrossRef]

15.   TAPPI. *Alpha-, Beta- and Gamma-Cellulose in Pulp*; Test Methods T 203 C; Technical Association of the Pulp and Paper Industry: Atlanta, GA, USA, 1999.

16.   Saito, T.; Hirota, M.; Tamura, N.; Kimura, S.; Fukuzumi, H.; Heux, L.; Isogai, A. Individualization of nano-sized plant cellulose fibrils by direct surface carboxylation using TEMPO catalyst under neutral conditions. *Biomacromolecules* **2009**, *10*, 1992–1996. [CrossRef]

17.   Saito, T.; Isogai, A. TEMPO-mediated oxidation of native cellulose. The effect of oxidation conditions on chemical and crystal structures of the water-insoluble fractions. *Biomacromolecules* **2004**, *5*, 1983–1989. [CrossRef] [PubMed]

18.   TAPPI. *Acid Insoluble Lignin in Wood and Pulp*; TAPPI Press: Atlanta, GA, USA, 2011.

19.   Segal, L.; Creely, J.J.; Martin, A.E.; Conrad, C.M. An Empirical Method for Estimating the Degree of Crystallinity of Native Cellulose Using the X-ray Diffractometer. *Text. Res. J.* **1959**, *29*, 786–794. [CrossRef]

20.   Sehaqui, H.; Zhou, Q.; Ikkala, O.; Berglund, L.A. Strong and tough cellulose nanopaper with high specific surface area and porosity. *Biomacromolecules* **2011**, *12*, 3638–3644. [CrossRef] [PubMed]

21.   Iwamoto, S.; Lee, S.-H.; Endo, T. Relationship between aspect ratio and suspension viscosity of wood cellulose nanofibers. *Polym. J.* **2014**, *46*, 73–76. [CrossRef]

22.   Gharehkhani, S.; Zhang, Y.; Fatehi, P. Lignin-derived platform molecules through TEMPO catalytic oxidation strategies. *Prog. Energy Combust. Sci.* **2019**, *72*, 59–89. [CrossRef]

23.   Sjöström, E. *Wood Chemistry: Fundamentals and Applications*, 2nd ed.; Academic Press, Inc.: San Diego, CA, USA, 1993; pp. 232–283.

24.   Li, Q.; Renneckar, S. Supramolecular structure characterization of molecularly thin cellulose i nanoparticles. *Biomacromolecules* **2011**, *12*, 650–659. [CrossRef]

25.   Usov, I.; Nyström, G.; Adamcik, J.; Handschin, S.; Schütz, C.; Fall, A.; Bergström, L.; Mezzenga, R. Understanding nanocellulose chirality and structure-properties relationship at the single fibril level. *Nat. Commun.* **2015**, *6*. [CrossRef] [PubMed]

26.   Nyström, G.; Arcari, M.; Adamcik, J.; Usov, I.; Mezzenga, R. Nanocellulose Fragmentation Mechanisms and Inversion of Chirality from the Single Particle to the Cholesteric Phase. *ACS Nano* **2018**, *12*, 5141–5148. [CrossRef]

27.   Jonasson, S.; Bünder, A.; Das, O.; Niittylä, T.; Oksman, K. Comparison of tension wood and normal wood for oxidative nanofibrillation and network characteristics. *Cellulose* **2020**, *28*, 1085–1104. [CrossRef]

28.   Moser, C. Manufacturing and Characterisation of Cellulose Nanofibers. Ph.D. Thesis, Royal Institute of Technology, Stockholm, Sweden, 25 January 2018.

29.   Ramiah, M.V. Thermogravimetric and differential thermal analysis of cellulose, hemicellulose, and lignin. *J. Appl. Polym. Sci.* **1970**, *14*, 1323–1337. [CrossRef]

30.   Lichtenstein, K.; Lavoine, N. Toward a deeper understanding of the thermal degradation mechanism of nanocellulose. *Polym. Degrad. Stab.* **2017**, *146*, 53–60. [CrossRef]

31.   French, A.D.; Santiago Cintrón, M. Cellulose polymorphy, crystallite size, and the Segal Crystallinity Index. *Cellulose* **2013**, *20*, 583–588. [CrossRef]

32.   Daicho, K.; Saito, T.; Fujisawa, S.; Isogai, A. The Crystallinity of Nanocellulose: Dispersion-Induced Disordering of the Grain Boundary in Biologically Structured Cellulose. *ACS Appl. Nano Mater.* **2018**, *1*, 5774–5785. [CrossRef]

33.   Meng, Q.; Wang, T.J. Mechanics of Strong and Tough Cellulose Nanopaper. *Appl. Mech. Rev.* **2019**, *71*. [CrossRef]

34.   Sehaqui, H.; Salajková, M.; Zhou, Q.; Berglund, L.A. Mechanical performance tailoring of tough ultra-high porosity foams prepared from cellulose i nanofiber suspensions. *Soft Matter* **2010**, *6*, 1824–1832. [CrossRef]

35.   Nge, T.T.; Lee, S.H.; Endo, T. Preparation of nanoscale cellulose materials with different morphologies by mechanical treatments and their characterization. *Cellulose* **2013**, *20*, 1841–1852. [CrossRef]

36.   Ding, Q.; Zeng, J.; Wang, B.; Tang, D.; Chen, K.; Gao, W. Effect of nanocellulose fiber hornification on water fraction characteristics and hydroxyl accessibility during dehydration. *Carbohydr. Polym.* **2019**, *207*, 44–51. [CrossRef]

nanomaterials

MDPI

Article

Ice-Templated Cellulose Nanofiber Filaments as a Reinforcement Material in Epoxy Composites

Tuukka Nissilä [1], Jiayuan Wei [2], Shiyu Geng [2], Anita Teleman [3] and Kristiina Oksman [2,4,*]

[1] Fibre and Particle Engineering Research Unit, Faculty of Technology, University of Oulu, FI-90014 Oulu, Finland; Tuukka.nissila@oulu.fi

[2] Division of Materials Science, Department of Engineering Sciences and Mathematics, Luleå University of Technology, SE-97187 Luleå, Sweden; Jiayuan.wei@ltu.se (J.W.); shiyu.geng@ltu.se (S.G.)

[3] RISE Research Institutes of Sweden, SE-11428 Stockholm, Sweden; anita.teleman@rise.se

[4] Mechanical & Industrial Engineering (MIE), University of Toronto, Toronto, ON M5S 3G8, Canada

* Correspondence: kristiina.oksman@ltu.se

Abstract: Finding renewable alternatives to the commonly used reinforcement materials in composites is attracting a significant amount of research interest. Nanocellulose is a promising candidate owing to its wide availability and favorable properties such as high Young's modulus. This study addressed the major problems inherent to cellulose nanocomposites, namely, controlling the fiber structure and obtaining a sufficient interfacial adhesion between nanocellulose and a non-hydrophilic matrix. Unidirectionally aligned cellulose nanofiber filament mats were obtained via ice-templating, and chemical vapor deposition was used to cover the filament surfaces with an aminosilane before impregnating the mats with a bio-epoxy resin. The process resulted in cellulose nanocomposites with an oriented structure and a strong fiber–matrix interface. Diffuse reflectance infrared Fourier transform and X-ray photoelectron spectroscopy studies revealed the presence of silane on the filaments. The improved interface, resulting from the surface treatment, was observable in electron microscopy images and was further confirmed by the significant increase in the tan delta peak temperature. The storage modulus of the matrix could be improved up to 2.5-fold with 18 wt% filament content and was significantly higher in the filament direction. Wide-angle X-ray scattering was used to study the orientation of cellulose nanofibers in the filament mats and the composites, and the corresponding orientation indices were 0.6 and 0.53, respectively, indicating a significant level of alignment.

Keywords: cellulose nanocomposite; ice-templating; interface; orientation; mechanical properties

Citation: Nissilä, T.; Wei, J.; Geng, S.; Teleman, A.; Oksman, K. Ice-Templated Cellulose Nanofiber Filaments as a Reinforcement Material in Epoxy Composites. *Nanomaterials* **2021**, *11*, 490. https://doi.org/10.3390/nano11020490

Academic Editors: Carla Vilela and Carmen S.R. Freire

Received: 21 January 2021
Accepted: 9 February 2021
Published: 15 February 2021

Publisher's Note: MDPI stays neutral with regard to jurisdictional claims in published maps and institutional affiliations.

1. Introduction

Nanocellulose is considered to be an ideal renewable alternative to the more commonly used reinforcement materials in composites [1]. It is widely available as it can be either produced from a variety of plant sources or synthesized by bacteria, and its mechanical properties are outstanding. The nanoscale fibers are more defect-free compared with natural fibers or microscopic native cellulose, and Young's modulus has been estimated to be approximately 100 GPa [2], a value comparable to that of Kevlar [3]. The strength can be as high as several gigapascals [1]. The nanoscale size also contributes to the additional benefit of high specific surface area [4,5], which leads to a high amount of interfacial area between the fibers and the matrix material and thus to a potentially strong reinforcement effect.

One of the main challenges in cellulose nanocomposite processing is controlling the fiber arrangement inside the matrix material, especially when hydrophobic polymers are used. A widely utilized method is creating various continuous structures, such as filaments, nanopapers, and aerogels. The tendency of nanocellulose to form networks via strong fiber–fiber hydrogen bonds has been known since the earliest studies on cellulose nanocomposites [6,7]. It was shown that the mechanical properties of a polymer matrix could be significantly improved even with a low fraction of the nanoscale fibers. The remarkable effect was attributed to a percolation phenomenon, i.e., the formation of a continuous network structure that provides a stress transfer mechanism outperforming the theoretical predictions for the mechanical properties of short-fiber composites. Later, the network-forming tendency of nanocellulose has been used to prepare nanopapers that have then been impregnated with a polymeric resin to obtain composites [8–10]. The high specific surface area reported for nanopapers suggests that such fiber networks consist of well-individualized nanoscale fibers that provide effective reinforcement [4]. This approach has resulted in relatively high fiber contents and promising mechanical properties. However, the filling process is typically time-consuming, and often an organic solvent has to be used to facilitate the impregnation. Another means of controlling the organization of the nanocellulosic entities is the formation of single filaments similar to those traditionally used in the textile industry [11–17]. Dry spinning, wet spinning, and hydrodynamic alignment are some variations of the same approach and have all been successfully utilized in nanocellulose processing. The filaments obtained via the spinning process are commonly in the micrometer scale. Considering the application of such materials in composites, the reported average diameters ranging 6.8–250 μm [11–17] suggest that the full potential of having a nanoscale raw material has not yet been reached. A better reinforcement effect could be expected by finding ways to decrease the lateral dimension of the filaments.

Ice-templating is a method that can be used to turn a nanocellulose water suspension into a porous honeycomb-like structure [18–20]. The resulting aerogels are composed of aligned pores that run through the material in the direction of the ice crystal growth and can be used as preforms for composite materials by filling the structure with a polymer [21–23]. However, these kinds of honeycomb structures can only be impregnated in the pore direction, making the process time-consuming and impractical. In addition, the relatively low specific surface area of ice-templated aerogels suggests that the nanofibers have aggregated during the formation of the honeycomb, and again, the benefit of having a nanoreinforcement material is lost [18,22,24]. Threadlike nanocellulose filaments can be produced with the same approach by decreasing the fiber content of the suspension [25,26]. Instead of forming a self-standing monolithic honeycomb foam, the nanofibers assemble into thin filaments with a diameter as small as a few hundred nanometers. In turn, the filaments are arranged into a mat-like and partially interconnected material in which they are oriented along the freezing direction. This type of open structure can be impregnated with a liquid resin in the through-plane direction instead of filling the honeycomb pores gradually from one end to the other.

In addition to the difficulty of controlling the arrangement of the nanocellulose fibers, the interfacial adhesion between the hydrophilic reinforcement material and the most common thermoset resins tends to be poor [27]. Using an organosilane to treat the surface of the fibers improves the properties of cellulose and natural fiber/epoxy composites [28,29]. This is the case especially when an aminosilane, capable of forming covalent bonds with the resin, is used [29]. However, the commonly used solution-based silylation is not easily applicable to dried nanocellulose materials, because they tend to deform during the evaporation of the water-containing solvent. Chemical vapor deposition (CVD) is a more straightforward and efficient way to functionalize and hydrophobize cellulosic nanomaterials, such as cellulose nanofiber aerogels [23] and filaments [16]. In its simplest manifestation, the process consists of placing the cellulosic material and the liquid silane inside a closed container and vaporizing the chemical by heat and/or vacuum to cover the cellulosic surfaces. Any solvents or subsequent processing steps are not needed.

In order to control the assembly of the cellulose nanofibers into a macroscopic structure, we utilized ice-templating to prepare oriented filament networks (Figure 1). These nonwoven cellulose nanofiber (CNF) filaments were then used as a reinforcement material in composites by filling the preforms with a bio-epoxy resin via vacuum infusion. The filament surfaces were treated with an aminosilane using a simple lab-scale CVD process to improve the interfacial adhesion with the matrix material. This paper presents the method of preparing the nanocellulose filaments and the composites along with the results of microscopy, diffuse reflectance infrared Fourier transform (DRIFT) spectroscopy, X-ray photoelectron spectroscopy (XPS), wide-angle X-ray scattering (WAXS), and dynamic mechanical analysis (DMA) studies.

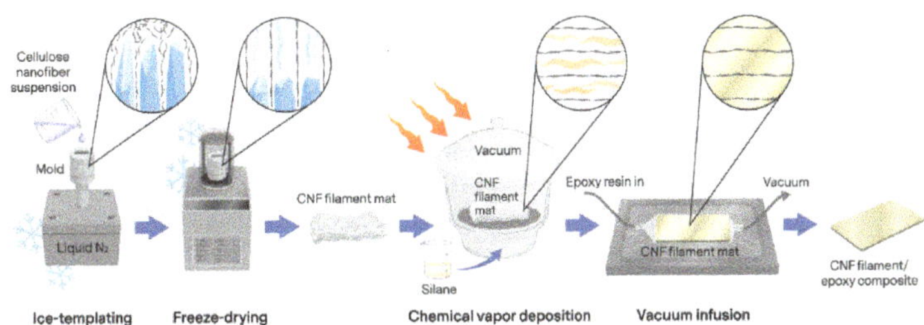

Figure 1. Process schematic. Ice-templating was used to orient the cellulose nanofibers (CNFs) into thin filaments, and dry filament mats were obtained after freeze-drying. The filament surfaces were covered with an aminosilane, and the treated filament mats were impregnated with a bio-epoxy resin to form composites.

2. Materials and Methods

2.1. Materials

Bleached softwood sulfate pulp (Stora Enso, Oulu, Finland) was used as a CNF source. The chemical composition of the pulp was reported in our previous study: cellulose 96.3%, hemicellulose 2.4%, lignin 1.3%, and the crystallinity index 68% [30]. The pulp with a 2.0 wt% concentration was passed through an ultrafine grinder (Supermasscolloider MKCA 6-2J CE, Masuko Sangyo Co., Ltd., Kawaguchi, Japan) until a final gap of −90 μm relative to the initial contact point of the grinding stones was reached. Further fibrillation was achieved via microfluidization (Microfluidics M-110EH-30, Westwood, MA, USA). The material was passed twice through 200 and 200 μm chambers at 1000 bar, twice through 200 and 100 μm chambers at 1500 bar, and finally twice through 200 and 87 μm chambers at 1500 bar. Figure 2 shows field emission scanning electron microscope (FESEM) and transmission electron microscope (TEM) images of the resulting CNFs. The size distribution was calculated based on measuring the width of approximately 250 fibers from the TEM images. The width distribution was between 10 and 30 nm. Super Sap CLR (Entropy Resins, Hayward, CA, USA) bio-epoxy was used as the polymer matrix. The resin was mixed with Super Sap INH and CLX hardeners using a 100:19:19 (CLR:INH:CLX) mixing ratio based on weight. 3-Aminopropyltriethoxysilane (APTES, 98%, Sigma-Aldrich, St. Louis, MO, USA) was used as received.

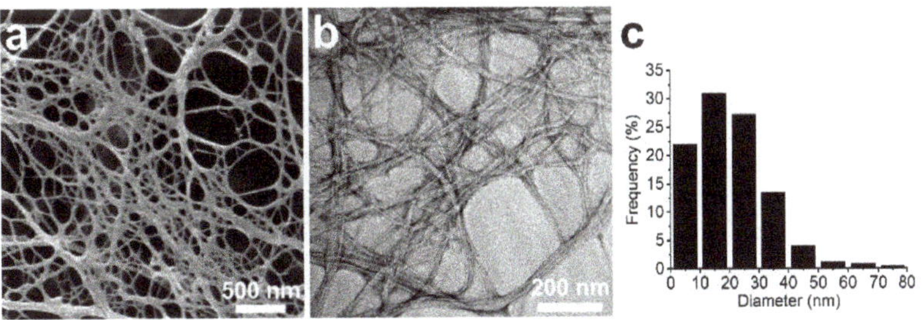

Figure 2. Cellulose nanofibers. (**a**) FESEM and (**b**) TEM images, and (**c**) the size distribution of the CNFs used as a raw material.

2.2. CNF Filament Preparation

CNF filaments were prepared via ice-templating using a setup described elsewhere [27]. The procedure consisted of pouring a 0.05 or 0.1 wt% CNF water suspension inside a polytetrafluoroethylene (PTFE) mold and unidirectionally freezing it. The copper bottom plate of the mold was placed on top of a copper rod immersed in liquid nitrogen. The temperature was controlled with a PID-controlled heating element attached to the rod. A cooling rate of 40 °C/h was used in all the experiments. Dry CNF filament mats were obtained by sublimating the ice inside a freeze dryer for approximately 4 days.

Silane-treated CNF filaments were prepared via CVD. A beaker with 1 mL of APTES and the freeze-dried filament mats were placed inside a desiccator. A vacuum pressure of approximately 0.95 bar was applied to the desiccator, which was then kept inside an oven at 150 °C for 1 h. After repressurizing and cooling the setup, CNF filament mats with an APTES surface were obtained.

2.3. Preparation of CNF Filament Composites

Composite materials were processed using vacuum infusion [22,27]. Four 0.05 wt% or two 0.1 wt% CNF filament mats were placed on a metal mold and covered with peel ply and breather cloth to facilitate the resin flow into the mat. Sealant tape and a plastic film were used to seal the system, and a vacuum was used to fill the filament mats with a degassed bio-epoxy resin. After filling the whole system with the resin, the outlet and inlet tubes were clamped, and the system was left to cure at room temperature. After 24 h, the samples were demolded, post-cured at 80 °C for 2 h, and left to cool to room temperature. Finally, the samples were polished to remove the surface roughness created by the peel ply.

2.4. Characterization

The CNF filaments were imaged with an optical microscope (Leica MZ FL III, Leica Camera AG, Wetzlar, Germany). A piece of the filament mat was placed between glass slides for imaging.

The CNFs, CNF filament mats, and composite fracture surfaces were imaged with FESEM (ZEISS ULTRA Plus FE-SEM, Carl Zeiss AG, Oberkochen, Germany). The CNF sample was carefully collected on a 0.2 μm polycarbonate membrane via vacuum filtration, frozen with liquid nitrogen, and freeze-dried. A piece of the membrane containing the dried CNFs was attached to a sample holder with carbon tape. The CNF filament samples were prepared by gently putting the filament mats in contact with carbon tape attached to sample holders, after which some of the filaments had been glued to the tape. The average diameters of the filaments were calculated by measuring 100 filaments from the SEM images. The composite fracture surfaces were prepared by immersing a piece of the material in liquid nitrogen and breaking it with two tweezers.

TEM (JEOL JEM-2200FS, JEOL Ltd., Tokyo, Japan) was used to image the CNFs. A drop of dilute suspension was applied on a carbon-coated grid and colored with uranyl acetate.

The specific surface area of the CNF mats was determined by N2 physisorption using the Brunauer–Emmett–Teller (BET) method (Micromeritics ASAP 2020). The samples were kept inside an oven at 105 °C for 18 h before testing to remove adsorbed moisture.

Viscoelastic properties of the composites were characterized with DMA (Q800 DMA, TA Instruments, New Castle, DE, USA). Specimens of size 30×3 mm^2 were cut from the vacuum-infused samples, polished to a thickness of 0.1 mm, and tested under tension mode. The span length was set at 15 mm. A displacement of 15 μm and a ramp rate of 2 °C/min from 30 to 150 °C were used. Only the 30–100 °C range is reported as no meaningful changes in the properties were detected at higher temperatures. The results were analyzed using one-way ANOVA and Tukey's tests with a 0.05 significance level.

DRIFT spectroscopy (Bruker Vertex 80V, Bruker, Billerica, MA, USA) was used to study the chemical composition of the CNF filament mats. The mats were pressed into pellets, and spectra from 400 to 4400 cm^{-1} were obtained. Forty scans with a 4 cm^{-1} resolution were performed for each sample.

XPS (ESCALAB 250Xi, Thermo Fisher Scientific, Loughborough, UK) was used to analyze the surface chemistry of the CNF filaments. A monochromatic Al Kα X-ray source was used at 300 W. The analyzer pass energy was 150 eV for the survey scan and 20 eV for the high-resolution scans.

The alignment of CNFs in the ice-templated CNF filament mats was studied by WAXS measurements on an Anton Paar SAXSpoint 2.0 system (Anton Paar, Graz, Austria) equipped with a Microsource X-ray source (Cu K-alpha radiation, wavelength 0.15418 nm) and a Dectris 2D CMOS Eiger R 1M detector with a 75×75 μm^2 pixel size. All measurements were performed with a beam size of approximately 500 μm diameter and a beam path pressure of about 1–2 mbar. The sample-to-detector distance was 111 mm during the measurements. All samples were mounted on a Sampler for Solids 10×10 mm^2 (Anton Paar, Graz, Austria) holder. Three frames of 24 min duration were read from the detector, giving a total measurement time of 1.2 h per sample. The transmittance was determined and used for scaling of intensities. The software used for instrument control was SAXSdrive version 2.01.224 (Anton Paar, Graz, Austria), and post-acquisition data processing was performed using the SAXSanalysis version 4.00.046 (Anton Paar, Graz, Austria).

The orientation of the cellulose crystals in the composites was studied at the beamline NanoMAX of the MAX IV synchrotron laboratory (Lund, Sweden), and the two-dimensional WAXS patterns were obtained. A photon energy of 10 keV was used. The beam size was 250×250 nm^2, and a sample area of 80×40 μm^2 was analyzed in a single scan. The orientation index (f_c) of the cellulose crystals was calculated according to the intensity distributions of the azimuthal angle using the following equation [31]:

$$f_c = \frac{180° - \mathrm{FWHM}}{180°},\tag{1}$$

where FWHM is the full width at the half-maximum of the azimuthal angle distribution.

3. Results and Discussion

3.1. CNF Filament Morphology

Figure 3a shows a photograph of an ice-templated CNF filament mat. The material is soft and does not retain the initial shape after demolding, unlike the aerogels reported previously [22,27]. The sample size is approximately $10 \times 6 \times 2$ cm^3 (length $\times$ width $\times$ thickness) when inside the mold but changes immediately during and after demolding. This is a result of the open filamentous structure (Figure 3b), which significantly differs from the closed honeycomb structure usually found in CNF aerogels [18,19,22,27]. Due to the low concentration, the CNFs in the water suspension have been arranged by the growing ice crystals into thin strands oriented in the freezing direction instead of forming hexagonal pores. Figure 3c–f shows the morphology of the filaments in detail. The average diameters of the filaments prepared from 0.05 and 0.1 wt% suspensions are 558 ± 186 and 1073 ± 472 nm, respectively, as measured from FESEM images, and the materials are henceforward called 0.56 and 1.1 μm filaments accordingly. The specific surface areas of the 0.56 and 1.1 μm CNF filament mats are 10.68 and 6.89 m^2/g, corresponding to theoretical average filament diameters of 250 and 387 nm. The difference between the measured and theoretical values might be due to the irregular shape and surface roughness that increase the surface areas of the filaments compared to perfectly smooth cylinders assumed in the theoretical values. In addition, some smaller filaments may have been unintentionally excluded from the manual measurements because they are less visible in FESEM images.

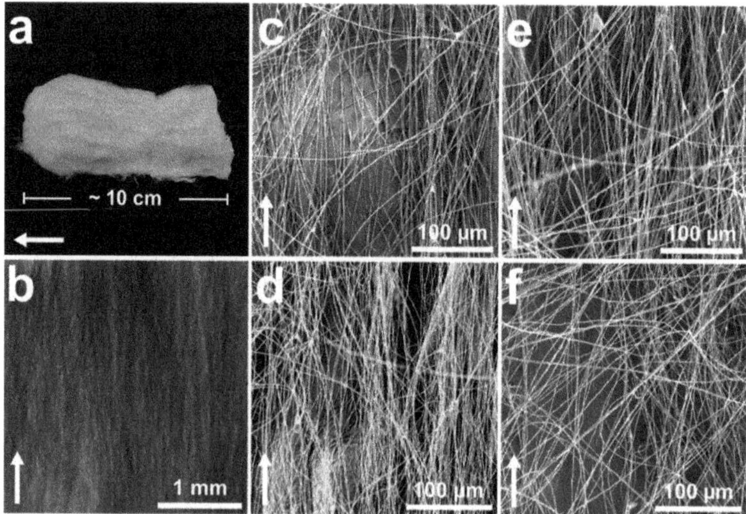

Figure 3. CNF filaments. (**a**) A photograph and (**b**) an optical microscope image of a CNF filament network. FESEM images of non-silylated (**c**) 1.1 and (**d**) 0.56 μm and silylated (**e**) 1.1 and (**f**) 0.56 μm filaments. The white arrow indicates the freezing direction.

The filaments are similar to those reported elsewhere. For example, Han et al. (2013) obtained oriented filaments with average diameters ranging from 0.57 to 1.5 μm when using a 0.05 wt% nanocellulose water suspension [26]. The diameter of the filaments was determined by the type of nanocellulose used. For cellulose nanocrystals (CNCs), the presence of surface sulfate groups, and thus greater repulsion between the fibers, was suggested to result in thinner filaments. The higher tendency of mechanically fibrillated CNFs to aggregate caused the corresponding filaments to have an average diameter twice as large as that of the nanocrystal-based ones. Meanwhile, Chen et al. (2014) reported diameters ranging from 50 to 300 nm and from 150 to 900 nm for filaments made from high-intensity ultrasonication-induced CNFs prepared with and without prior TEMPO-mediated oxidation, respectively [25]. The authors did not elaborate why the sizes are different, but it can be deduced that the electrostatic repulsion caused by the carboxyl groups on the surface of TEMPO-CNFs, with the additional influence of the reported smaller nanofiber size, resulted in a smaller filament diameter. The size of the filaments reported here falls on the upper end of the size range for similarly prepared materials, agreeing well with the fact that mechanically fibrillated CNFs with no surface functionality were used as a raw material. A reduction in the average diameter could be expected by utilizing, for example, CNCs or TEMPO-CNFs as a raw material.

3.2. CNF Orientation in the Filaments

Figure 4 shows a WAXS image together with an identification of the cellulose crystal structure and distribution of orientation for crystalline cellulose. The (200), (110), and (110) reflections can be used to quantify the orientation of both the cellulose crystals and the CNFs since the crystals are aligned in the direction of the nanofibers. By determining the baseline and maximum intensities in the intensity distribution of the azimuthal angle of the (110) and (110) reflections, the orientation index f_c of the CNFs in the filaments can be calculated according to Equation (1). If all nanofibers are aligned in the same direction, $f_c = 1$, and if they are randomly distributed, $f_c = 0$. The orientation index for the ice-templated CNF filaments prepared from both 0.05 and 0.1 wt% CNF water suspensions is 0.6, indicating that the nanofibers are partially oriented along the filaments. It should be noted that the calculated orientation index is related to the orientation of the CNFs in a bundle of filaments and is not directly related to the orientation of the microscopic filaments inside the bundle or to the orientation of the CNFs in single filaments.

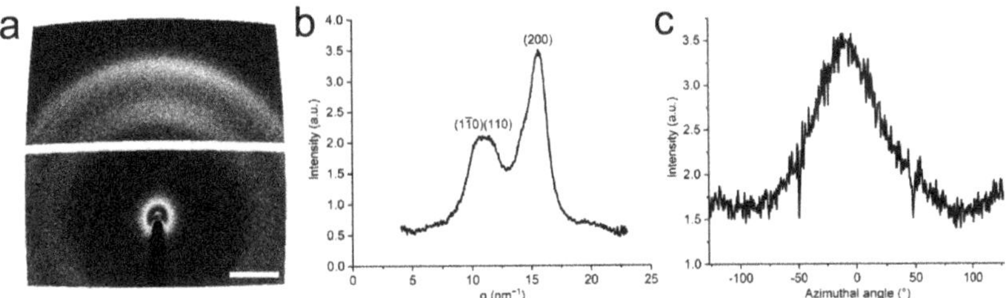

Figure 4. Cellulose crystal orientation in the filaments. (**a**) A representative WAXS diffractogram of a CNF filament in the bundle. The scale bar represents $q = 5\,\mathrm{nm}^{-1}$. Sample direction is set on equatorial direction. Sample and nanofiber directions coincide. (**b**) Radial integration of the diffractogram after background subtraction. (**c**) Azimuthal integration of the $(1\bar{1}0)$ and (110) scattering planes, $q = 11 \pm 0.5\,\mathrm{nm}^{-1}$. The zero degree of azimuthal angle is set on meridian.

Orientation index values ranging from approximately 0.56 to 0.82 have been reported for ice-templated nanocellulose networks [18,24]. The value was almost constant at higher fiber contents but dropped dramatically when the suspension concentration was lower than a certain critical value [18]. This critical value was 0.2 wt% for CNCs and 0.08 wt% for CNFs. The drop in the orientation index was suggested to be caused by wrinkling and bending of the fiber structures in the low fiber content networks. Similarly, the filaments reported in the current study are soft, causing the structure to change during demolding and handling of the samples. The orientation of the filaments, and thus also the CNFs, in the dried filament bundles is not the same as the orientation after the freezing process and prior to drying. The dried structure is not fixed but gets easily collapsed, and the filaments do not retain their original alignment, as seen in Figure 3a.

Orientation indices ranging from 0.83 to 0.92 have been reported for CNF filaments prepared via flow focusing [17]. However, the measurements were conducted on single filaments, making the CNF orientation a property of the filament itself and not that of an assembly of filaments as is the case in the current study. The filaments reported here are not readily separable from the network they are a part of, and no information on the nanofiber orientation inside single filaments can be currently provided. Another approach is inducing orientation in cellulose nanopapers by mechanically pulling, or drawing, the wet papers [32]. A highest orientation index of 0.82 was reported for such CNF networks. However, these kinds of networks are difficult to impregnate, as discussed in the introduction, and, unlike the filament mats presented here, are not easily applicable in traditional composite processing.

3.3. CNF Filament Surface Characteristics

The CNF filaments were analyzed using DRIFT and XPS to evaluate the CVD process and the resulting aminosilane surface coverage (Figure 5). The DRIFT spectra (Figure 5a) show typical cellulose peaks at around 3380 and 1640 cm^{-1}, 2900 and 1250–1460 cm^{-1}, and 1050–1170 cm^{-1}, which can be attributed to hydroxyl (OH), alkyl (CH_2), and C–O–C groups, respectively [33,34], and are observed in all samples. On the other hand, the peak at 1570 cm^{-1} can be assigned to N–H bending of the primary amine of the silane molecule [34–37] and is only present in the silylated samples. This is a clear indication of a successful surface coverage. The potential Si–O bridges are indistinguishable from the C–O–C vibrations, and no conclusions can be made concerning the bonding between the aminosilane and cellulose molecules [16,38]. However, APTES has been shown to form covalent bonds with cellulose when treated at an elevated temperature [39].

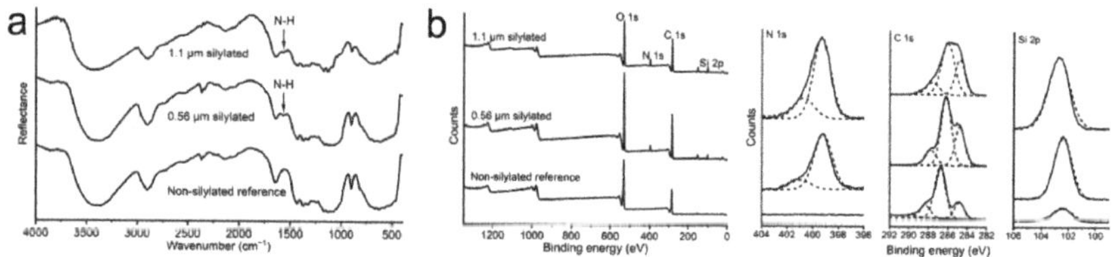

Figure 5. DRIFT and XPS spectra. (**a**) DRIFT and (**b**) XPS spectra of the CNF filament mats.

Figure 5b shows the XPS spectra of a reference non-silylated filament mat and silylated 0.05 and 0.1 wt% filament mats. All samples show the typical oxygen (O 1s) and carbon (C 1s) peaks at 533 and 286 eV, respectively [40]. Both of the silylated samples show two additional peaks at 399 and 102 eV, corresponding to nitrogen (N 1s) and silicon (Si 2p), respectively [40]. These peaks indicate the presence of the aminosilane chemical on the filament surface, further confirming the results from DRIFT. The reference sample also shows a minor silicon peak, which is most likely caused by a contamination originating from the various processing steps such as the mechanical grinding.

The C 1s peak is divided into several components for all samples. The peaks at 288.1 and 286.6 eV can be ascribed to O–C–O or C=O and C–O bonds, respectively, and are typical of cellulosic materials [16,41]. The peak at 284.8 eV is related to both C–C and C–Si bonds. The relative size of this peak is significantly bigger for the silylated samples, further confirming the presence of the aminosilane molecule, which has a backbone consisting of three carbon atoms and one silicon atom (Table 1).

Table 1. Chemical composition of the CNF filament surfaces. Fractions of elements in percentages (%) on the silylated and non-silylated filament surfaces according to XPS analysis.

Sample	O 1s	C 1s	N 1s	Si 2p	(C–C + C–Si): (O–C–O + C–O)
1.1 μm silylated	30.05	55.31	6.55	7.64	0.42
0.56 μm silylated	40.39	50.85	3.88	4.89	0.44
Non-silylated reference	42.58	55.54	0.00	1.78	0.21

3.4. Composite Morphology

The composite fracture surfaces show the CNF filaments embedded in the epoxy matrix (Figure 6). The side profiles of the freezing-oriented filaments are seen in the longitudinal cross sections (Figure 6a). A significant difference can be observed between the silylated and the non-silylated samples. In the non-silylated composites, the filaments appear to be separate from the matrix with visible gaps between the two components. On the other hand, the silylated filaments are well integrated in the matrix. This is even more pronounced in the transverse cross sections (Figure 6b). The non-silylated samples show a substantial number of fiber pullouts, and there appear to be significant gaps between the filaments and the matrix, indicating poor interfacial adhesion. Similar findings were made in a previous study [27]. The silylated filaments form a more homogenous material with the epoxy matrix, and the breakage has primarily occurred in a brittle fashion in contrast to debonding.

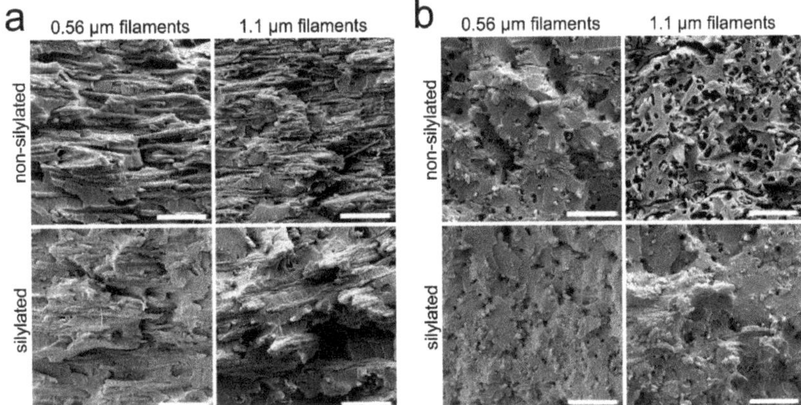

Figure 6. Composite morphology. Longitudinal (**a**) and transverse (**b**) fracture surfaces of the CNF filament/epoxy composites. (The scale bar is 5 µm in all images.)

3.5. CNF Orientation in the Composites

The orientation of cellulose crystal and the CNFs in the composites were also analyzed, and Figure 7a,b shows a representative 2D-WAXS diffractogram of a 1.1 µm filament/epoxy composite and its corresponding radial integration, respectively. The main peak at 2θ of 18.8°, attributed to the presence of epoxy [42], merges with the strongest cellulose peak from (200) planes seen at 22.5°, making it difficult to analyze the CNF alignment. However, the peak from cellulose (004) planes at 34.3° is not affected by epoxy [43]. An azimuthal integration of the (004) planes was carried out, and the resulting curve is shown in Figure 7c. The orientation index calculated from Equation (1) is 0.53, indicating that the CNF alignment is retained after the epoxy impregnation.

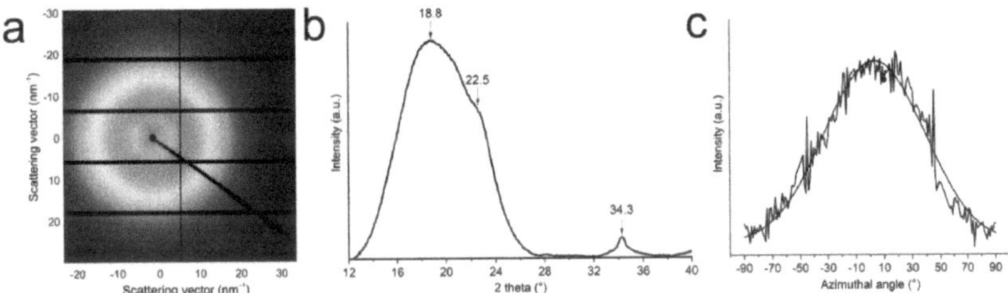

Figure 7. Analysis of filament alignment in the composites using 2D-WAXS. The meridional direction coincides with the freezing direction during ice-templating, i.e., the filament direction. (**a**) A representative diffractogram of a 1.1 µm filament/epoxy composite. (**b**) Radial integration of the diffractogram after baseline subtraction. (**c**) Azimuthal integration of the (004) plane. The zero degree was set on meridian.

The orientation index value is lower than the 0.84 reported for cellulose nanocomposites prepared from CNCs and carboxymethyl cellulose [44]. However, this value is not directly comparable to the results obtained in the current study, as the orientation was induced by drawing a water-based mixture of nanofibers and a polymer matrix. The method is not applicable to thermoset composites. To the best of our knowledge, quantitative data on the orientation of CNFs, or any other type of nanocellulose, in thermoset composites have not been reported before.

3.6. Composite Viscoelastic Properties

All of the composites with 18 wt% CNF filament content show improved storage modulus values throughout the 30–100 °C test range (Figure 8, Table 2). The improvement is most significant in the rubbery state due to a continuous fiber network formed by the filaments, a phenomenon widely reported for cellulose nanocomposites [6,9,27,45]. The storage modulus in the longitudinal, i.e., filament, direction is increased up to 2.5-fold compared to that of the epoxy polymer at 30 °C. Thus, the reinforcement effect is more significant than in the previously reported aerogel-based epoxy composites [22,27] and is comparable to those reported for epoxy composites with 20 wt% tunicate whiskers (3.1-fold) [45]. On the other hand, the highest storage modulus value in the transverse direction corresponds to a 1.7-fold increase. All samples show higher values in the longitudinal direction resulting from the freezing process used for preparing the filament mats. The vertically advancing ice front has pushed the suspended CNFs into threadlike filaments that are oriented along the freezing direction. However, it should be emphasized that the difference in the longitudinal and transverse properties is not directly related to the orientation of individual CNFs but to that of the filaments formed by the nanofibers. Only by combining the observations from the WAXS and DMA studies, it can be deduced that both the single CNFs and the filaments formed by the CNFs are at least partially oriented in the composites and that these two things are related to each other.

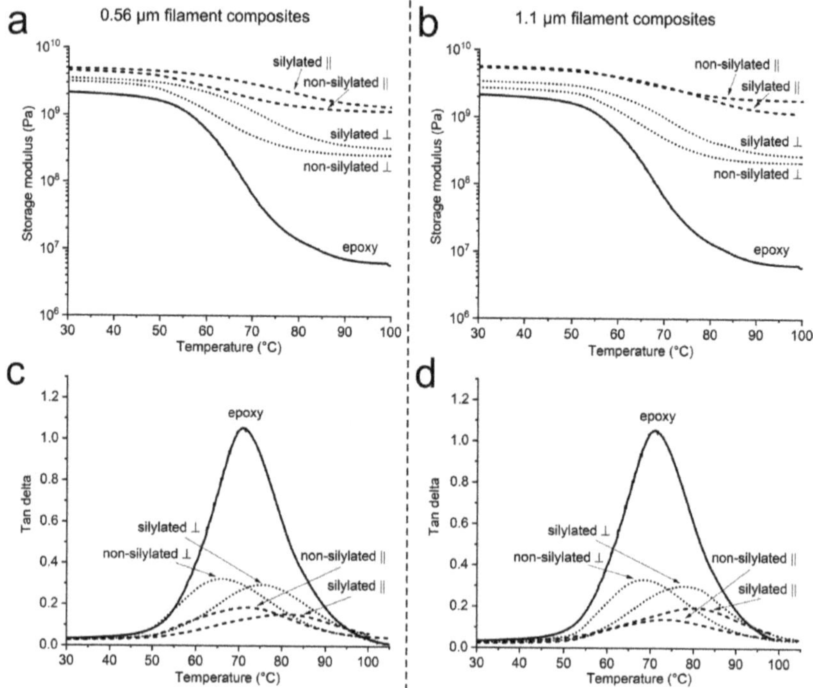

Figure 8. Viscoelastic properties of the CNF filament/epoxy composites. Storage modulus of the (**a**) 0.56 and (**b**) 1.1 μm filament composites. Tan delta of the (**c**) 0.56 and (**d**) 1.1 μm filament composites. Longitudinal (‖) and transverse (⊥) specimens were prepared from the composite samples.

Table 2. Viscoelastic properties of the CNF filament/epoxy composites. Longitudinal ($\parallel$) and transverse ($\perp$) specimens were prepared from the composite samples.

Sample	Storage Modulus at 30 °C (MPa) *	Storage Modulus at 100 °C (MPa) *	Tan Delta Peak (°C) *
Epoxy	2180 ($\pm$ 180) [a]	6.52 ($\pm$ 0.53) [a]	72.1 ($\pm$ 1.0) [a]
0.56 μm $\parallel$	5030 ($\pm$ 510) [b]	1270 ($\pm$ 230) [b]	71.3 ($\pm$ 1.1) [ab]
0.56 μm silylated $\parallel$	5050 ($\pm$ 360) [b]	1250 ($\pm$ 100) [b]	80.2 ($\pm$ 0.4) [d]
0.56 μm $\perp$	3270 ($\pm$ 170) [a]	270 ($\pm$ 64) [a]	66.5 ($\pm$ 0.6) [c]
0.56 μm silylated $\perp$	3600 ($\pm$ 260) [a]	381 ($\pm$ 69) [a]	76.4 ($\pm$ 0.7) [e]
1.1 μm $\parallel$	5350 ($\pm$ 560) [b]	1440 ($\pm$ 440) [b]	73.4 ($\pm$ 0.5) [a]
1.1 μm silylated $\parallel$	5260 ($\pm$ 390) [b]	1070 ($\pm$ 170) [b]	80.2 ($\pm$ 0.9) [d]
1.1 μm $\perp$	2970 ($\pm$ 350) [a]	187 ($\pm$ 84) [a]	68.8 ($\pm$ 1.1) [bc]
1.1 silylated $\perp$	3360 ($\pm$ 140) [a]	282 ($\pm$ 91) [a]	78.2 ($\pm$ 0.5) [de]

* Values with the same superscript (a, b, c, d, and e) within a column do not differ significantly according to one-way analysis of variance (ANOVA) and Tukey's tests with a 0.05 significance level.

The silylation does not have a significant effect on the storage moduli. The storage modulus in the transverse direction is higher for the composites with silane-treated CNF filaments, but the difference is not significant, and the longitudinal values are almost identical. However, the silylation has a remarkable effect on the tan delta peak temperature. The peak temperature of the longitudinal specimens is shifted from 72.1 to 80.2 °C for both of the composites with silylated filaments. This indicates good interfacial adhesion between the filaments and the epoxy matrix and is a major improvement compared to the nontreated ones, showing no temperature shift at all. The values are also superior to those reported earlier for CNF aerogel/epoxy composites [22,27] and the increase in the tan delta peak temperature is comparable in scale to that achieved with cellulose whiskers using a solvent casting approach [45].

It is noteworthy that all samples show significant variation between the test specimens, indicating the presence of inhomogeneities in the fiber contents of the composites. The three test specimens were cut out from different parts of the samples and may thus contain different fiber fractions. In addition, the small specimen size and the fact that the samples were not perfectly even due to manual polishing may have contributed to errors in the measured specimen dimensions. Thus, not all of the differences in the DMA results are statistically significant (Table 2).

4. Conclusions

Trying to find ways of utilizing renewable raw materials such as cellulose instead of petroleum- or mineral-based ones in everyday applications is becoming more and more important. Composite materials are widely used as load-bearing structures, for example, in transportation, but the traditional reinforcement materials they contain are based on nonrenewable resources. Natural fibers from various plants can be used, but they do not necessarily meet the demands of many modern products. By processing the fibers into the smallest structural units, namely, nanocellulose, the remarkable mechanical and physical properties of the cellulose crystal can be accessed.

A major challenge in using nanocellulose in composites is controlling the micro- and macroscopic architecture and avoiding the fibers from forming agglomerates that negatively affect the properties of the final product. Ice-templating is a fascinating approach to assemble nano- and microscopic particles into macroscopic objects and applying the method to nanocellulose gives rise to lightweight foamlike materials in which the structure is oriented along the freezing direction of the ice crystals. With a low-enough fiber concentration, a network of interconnected thin filaments is formed instead of a honeycomb structure. In theory, the smaller the filament diameter, the better the properties of both the filaments and the composite material containing them. Thus, the possibility of obtaining ultrathin filaments with diameters less than 1 μm is intriguing.

This study shows the applicability of the ice-templating approach in preparing cellulose nanocomposites. Thin filaments were formed from mechanically processed CNFs, and the filament network structures were impregnated with a bio-epoxy resin using vacuum infusion. The resulting composites had an 18 wt% fiber content, and the mechanical properties were significantly improved compared to neat epoxy. The fiber–matrix interface could be optimized by utilizing a simple CVD process to cover the cellulosic surfaces with an aminosilane capable of forming covalent bonds with the epoxy polymer. The effect of the silylation procedure was distinguishable in both the FESEM and DMA studies. The treated filaments were well integrated into the matrix, and the tan delta peak temperature was up to 8 °C higher for the composites with silylated filaments.

Author Contributions: Conceptualization, K.O. and T.N.; methodology, K.O. and T.N.; validation, K.O., T.N., J.W., S.G., and A.T.; investigation, T.N., J.W., S.G., and A.T.; writing—original draft preparation, T.N., J.W., S.G., and A.T.; writing—review and editing, T.N., J.W., S.G, A.T., and K.O.; visualization, T.N., J.W., S.G., and A.T.; supervision, K.O.; project administration, K.O.; funding acquisition, K.O. All authors have read and agreed to the published version of the manuscript.

Funding: Business Finland (formerly the Finnish Funding Agency for Technology and Innovation, TEKES) is acknowledged for their financial support (grant no. 1841/31/2014). Part of the work was carried out with the support of the Centre for Material Analysis, University of Oulu, Finland. We acknowledge Bio4Energy project for financial support, MAX IV Laboratory for time on beamline NanoMAX under Proposal 20190363. Research conducted at MAX IV, a Swedish national user facility, is supported by the Swedish Research Council under contract 2018-07152, the Swedish Governmental Agency for Innovation Systems under contract 2018-04969, and Formas under contract 2019-02496. Treesearch Research Infrastructure is acknowledged for their financial support of the WAXS analysis at RISE.

Institutional Review Board Statement: Not applicable.

Informed Consent Statement: Not applicable.

Data Availability Statement: Data presented in this study are available on request from the corresponding author.

Acknowledgments: The authors would like to thank Kim Nygård and Ulf Johansson from MAX IV Laboratory for the help in conducting the WAXS experiment and the related data analysis, Stora Enso (Oulu, Finland) for providing the pulp raw material, and Jani Österlund for performing the nanofibrillation.

Conflicts of Interest: The authors declare no conflict of interest. The funders had no role in the design of the study; in the collection, analyses, or interpretation of data; in the writing of the manuscript; or in the decision to publish the results.

References

1. Lee, K.-Y.; Aitomäki, Y.; Berglund, L.A.; Oksman, K.; Bismarck, A. On the Use of Nanocellulose as Reinforcement in Polymer Matrix Composites. *Compos. Sci. Technol.* **2014**, *105*, 15–27. [CrossRef]
2. Dufresne, A. Nanocellulose: A New Ageless Bionanomaterial. *Mater. Today* **2013**, *16*, 220–227. [CrossRef]
3. Yeh, W.Y.; Young, R.J. Molecular Deformation Processes in Aromatic High Modulus Polymer Fibres. *Polymer* **1999**, *40*, 857–870. [CrossRef]
4. Sehaqui, H.; Zhou, Q.; Ikkala, O.; Berglund, L.A. Strong and Tough Cellulose Nanopaper with High Specific Surface Area and Porosity. *Biomacromolecules* **2011**, *12*, 3638–3644. [CrossRef] [PubMed]
5. Dufresne, A. *Nanocellulose: From Nature to High Performance Tailored Materials*; De Gruyter: Berlin, Germany; Boston, MA, USA, 2012. [CrossRef]
6. Favier, V.; Canova, G.R.; Cavaillé, J.Y.; Chanzy, H.; Dufresne, A.; Gauthier, C. Nanocomposite Materials from Latex and Cellulose Whiskers. *Polym. Adv. Technol.* **1995**, *6*, 351–355. [CrossRef]
7. Favier, V.; Chanzy, H.; Cavaille, J.Y. Polymer Nanocomposites Reinforced by Cellulose Whiskers. *Macromolecules* **1995**, *28*, 6365–6367. [CrossRef]
8. Nakagaito, A.N.; Yano, H. Novel High-Strength Biocomposites Based on Microfibrillated Cellulose Having Nano-Order-Unit Web-Like Network Structure. *Appl. Phys. A* **2005**, *80*, 155–159. [CrossRef]
9. Henriksson, M.; Fogelström, L.; Berglund, L.A.; Johansson, M.; Hult, A. Novel Nanocomposite Concept Based on Cross-Linking of Hyperbranched Polymers in Reactive Cellulose Nanopaper Templates. *Compos. Sci. Technol.* **2011**, *71*, 13–17. [CrossRef]

10. Aitomäki, Y.; Moreno-Rodriguez, S.; Lundström, T.S.; Oksman, K. Vacuum Infusion of Cellulose Nanofibre Network Composites: Influence of Porosity on Permeability and Impregnation. *Mater. Des.* **2016**, *95*, 204–211. [CrossRef]

11. Hooshmand, S.; Aitomäki, Y.; Norberg, N.; Mathew, A.P.; Oksman, K. Dry-Spun Single-Filament Fibers Comprising Solely Cellulose Nanofibers from Bioresidue. *ACS Appl. Mater. Interfaces* **2015**, *7*, 13022–13028. [CrossRef]

12. Hooshmand, S.; Aitomäki, Y.; Berglund, L.; Mathew, A.P.; Oksman, K. Enhanced Alignment and Mechanical Properties through the Use of Hydroxyethyl Cellulose in Solvent-Free Native Cellulose Spun Filaments. *Compos. Sci. Technol.* **2017**, *150*, 79–86. [CrossRef]

13. Iwamoto, S.; Isogai, A.; Iwata, T. Structure and Mechanical Properties of Wet-Spun Fibers Made from Natural Cellulose Nanofibers. *Biomacromolecules* **2011**, *12*, 831–836. [CrossRef]

14. Walther, A.; Timonen, J.V.I.; Díez, I.; Laukkanen, A.; Ikkala, O. Multifunctional High-Performance Biofibers Based on Wet-Extrusion of Renewable Native Cellulose Nanofibrils. *Adv. Mater.* **2011**, *23*, 2924–2928. [CrossRef]

15. Håkansson, K.M.O.; Fall, A.B.; Lundell, F.; Yu, S.; Krywka, C.; Roth, S.V.; Santoro, G.; Kvick, M.; Prahl Wittberg, L.; Wågberg, L.; et al. Hydrodynamic Alignment and Assembly of Nanofibrils Resulting in Strong Cellulose Filaments. *Nat. Commun.* **2014**, *5*, 4018. [CrossRef]

16. Cunha, A.G.; Lundahl, M.; Ansari, M.F.; Johansson, L.S.; Campbell, J.M.; Rojas, O.J. Surface Structuring and Water Interactions of Nanocellulose Filaments Modified with Organosilanes Toward Wearable Materials. *ACS Appl. Nano Mater.* **2018**, *1*, 5279–5288. [CrossRef]

17. Mittal, N.; Ansari, F.; Gowda, V.K.; Brouzet, C.; Chen, P.; Larsson, P.T.; Roth, S.V.; Lundell, F.; Wågberg, L.; Kotov, N.A.; et al. Multiscale Control of Nanocellulose Assembly: Transferring Remarkable Nanoscale Fibril Mechanics to Macroscale Fibers. *ACS Nano* **2018**, *12*, 6378–6388. [CrossRef]

18. Munier, P.; Gordeyeva, K.; Bergström, L.; Fall, A.B. Directional Freezing of Nanocellulose Dispersions Aligns the Rod-Like Particles and Produces Low-Density and Robust Particle Networks. *Biomacromolecules* **2016**, *17*, 1875–1881. [CrossRef]

19. Pan, Z.Z.; Nishihara, H.; Iwamura, S.; Sekiguchi, T.; Sato, A.; Isogai, A.; Kang, F.; Kyotani, T.; Yang, Q.H. Cellulose Nanofiber as a Distinct Structure-Directing Agent for Xylem-Like Microhoneycomb Monoliths by Unidirectional Freeze-Drying. *ACS Nano* **2016**, *10*, 10689–10697. [CrossRef] [PubMed]

20. Zeng, Z.; Wu, T.; Han, D.; Ren, Q.; Siqueira, G.; Nyström, G. Ultralight, Flexible, and Biomimetic Nanocellulose/Silver Nanowire Aerogels for Electromagnetic Interference Shielding. *ACS Nano* **2020**, *14*, 2927–2938. [CrossRef]

21. Barari, B.; Ellingham, T.K.; Qamhia, I.I.; Pillai, K.M.; El-Hajjar, R.; Turng, L.-S.; Sabo, R. Mechanical Characterization of Scalable Cellulose Nano-Fiber Based Composites Made Using Liquid Composite Molding Process. *Compos. Part B* **2016**, *84*, 277–284. [CrossRef]

22. Nissilä, T.; Karhula, S.S.; Saarakkala, S.; Oksman, K. Cellulose Nanofiber Aerogels Impregnated with Bio-Based Epoxy Using Vacuum Infusion: Structure, Orientation and Mechanical Properties. *Compos. Sci. Technol.* **2018**, *155*, 64–71. [CrossRef]

23. Zhai, T.; Zheng, Q.; Cai, Z.; Turng, L.S.; Xia, H.; Gong, S. Poly(Vinyl Alcohol)/Cellulose Nanofibril Hybrid Aerogels with an Aligned Microtubular Porous Structure and Their Composites with Polydimethylsiloxane. *Appl. Mater. Interfaces* **2015**, *7*, 7436–7444. [CrossRef] [PubMed]

24. Kriechbaum, K.; Munier, P.; Apostolopoulou-Kalkavoura, V.; Lavoine, N. Analysis of the Porous Architecture and Properties of Anisotropic Nanocellulose Foams: A Novel Approach to Assess the Quality of Cellulose Nanofibrils (CNFs). *ACS Sustain. Chem. Eng.* **2018**, *6*, 11959–11967. [CrossRef]

25. Chen, W.; Li, Q.; Wang, Y.; Yi, X.; Zeng, J.; Yu, H.; Liu, Y.; Li, J. Comparative Study of Aerogels Obtained from Differently Prepared Nanocellulose Fibers. *ChemSusChem* **2014**, *7*, 154–161. [CrossRef] [PubMed]

26. Han, J.; Zhou, C.; Wu, Y.; Liu, F.; Wu, Q. Self-Assembling Behavior of Cellulose Nanoparticles during Freeze-Drying: Effect of Suspension Concentration, Particle Size, Crystal Structure, and Surface Charge. *Biomacromolecules* **2013**, *14*, 1529–1540. [CrossRef] [PubMed]

27. Nissilä, T.; Hietala, M.; Oksman, K. A Method for Preparing Epoxy-Cellulose Nanofiber Composites with an Oriented Structure. *Compos. Part A* **2019**, *125*, 105515. [CrossRef]

28. George, J.; Ivens, J.; Verpoest, I. Mechanical Properties of Flax Fibre Reinforced Epoxy Composites. *Angew. Makromol. Chem.* **1999**, *272*, 41–45. [CrossRef]

29. Abdelmouleh, M.; Boufi, S.; Belgacem, M.N.; Dufresne, A.; Gandini, A. Modification of Cellulose Fibers with Functionalized Silanes: Effect of the Fiber Treatment on the Mechanical Performances of Cellulose-Thermoset Composites. *J. Appl. Polym. Sci.* **2005**, *98*, 974–984. [CrossRef]

30. Hietala, M.; Varrio, K.; Berglund, L.; Soini, J.; Oksman, K. Potential of municipal solid waste paper as raw material for production of cellulose nanofibres. *Waste Manag.* **2018**, *80*, 319–326. [CrossRef]

31. Lundahl, M.J.; Cunha, A.G.; Rojo, E.; Papageorgiou, A.C.; Rautkari, L.; Arboleda, J.C.; Rojas, O.J. Strength and Water Interactions of Cellulose I Filaments Wet-Spun from Cellulose Nanofibril Hydrogels. *Sci. Rep.* **2016**, *6*, 30695. [CrossRef]

32. Sehaqui, H.; Ezekiel Mushi, N.; Morimune, S.; Salajkova, M.; Nishino, T.; Berglund, L.A. Cellulose Nanofiber Orientation in Nanopaper and Nanocomposites by Cold Drawing. *ACS Appl. Mater. Interfaces* **2012**, *4*, 1043–1049. [CrossRef] [PubMed]

33. Qian, S.; Sheng, K.; Yu, K.; Xu, L.; Fontanillo Lopez, C.A. Improved Properties of PLA Biocomposites Toughened with Bamboo Cellulose Nanowhiskers through Silane Modification. *J. Mater. Sci.* **2018**, *53*, 10920–10932. [CrossRef]

34. Fernandes, S.C.M.; Sadocco, P.; Alonso-Varona, A.; Palomares, T.; Eceiza, A.; Silvestre, A.J.D.; Mondragon, I.; Freire, C.S.R. Bioinspired Antimicrobial and Biocompatible Bacterial Cellulose Membranes Obtained by Surface Functionalization with Aminoalkyl Groups. *ACS Appl. Mater. Interfaces* **2013**, *5*, 3290–3297. [CrossRef]
35. Selulosa-Polivinilklorida, S.R.N.; Sheltami, R.M.; Kargarzadeh, H.A.N.I.E.H.; Abdullah, I.B.R.A.H.I.M. Effects of Silane Surface Treatment of Cellulose Nanocrystals on the Tensile Properties of Cellulose-Polyvinyl Chloride Nanocomposite. *JSM* **2015**, *44*, 801–810. [CrossRef]
36. Pacaphol, K.; Aht-Ong, D. The Influences of Silanes on Interfacial Adhesion and Surface Properties of Nanocellulose Film Coating on Glass and Aluminum Substrates. *Surf. Coat. Technol.* **2017**, *320*, 70–81. [CrossRef]
37. Lu, J.; Askeland, P.; Drzal, L.T. Surface Modification of Microfibrillated Cellulose for Epoxy Composite Applications. *Polymer* **2008**, *49*, 1285–1296. [CrossRef]
38. Reale Batista, M.D.; Drzal, L.T. Carbon Fiber/Epoxy Matrix Composite Interphases Modified with Cellulose Nanocrystals. *Compos. Sci. Technol.* **2018**, *164*, 274–281. [CrossRef]
39. Abdelmouleh, M.; Boufi, S.; ben Salah, A.; Belgacem, M.N.; Gandini, A. Interaction of Silane Coupling Agents with Cellulose. *Langmuir* **2002**, *18*, 3203–3208. [CrossRef]
40. Khanjanzadeh, H.; Behrooz, R.; Bahramifar, N.; Gindl-Altmutter, W.; Bacher, M.; Edler, M.; Griesser, T. Surface Chemical Functionalization of Cellulose Nanocrystals by 3-Aminopropyltriethoxysilane. *Int. J. Biol. Macromol.* **2018**, *106*, 1288–1296. [CrossRef]
41. Kontturi, E.J. Surface Chemistry of Cellulose: From Natural Fibres to Model Surfaces. Doctoral thesis, Technische Universiteit Eindhoven, Eindhoven, The Netherlands, 2005.
42. Cui, M.; Ren, S.; Qin, S.; Xue, Q.; Zhao, H.; Wang, L. Non-Covalent Functionalized Hexagonal Boron Nitride Nanoplatelets to Improve Corrosion and Wear Resistance of Epoxy Coatings. *RSC Adv.* **2017**, *7*, 44043–44053. [CrossRef]
43. Wada, M.; Sugiyama, J.; Okano, T. Native Celluloses on the Basis of Two Crystalline Phase (Iα/Iβ) System. *J. Appl. Polym. Sci.* **1993**, *49*, 1491–1496. [CrossRef]
44. Wang, B.; Torres-Rendon, J.G.; Yu, J.; Zhang, Y.; Walther, A. Aligned Bioinspired Cellulose Nanocrystal-Based Nanocomposites with Synergetic Mechanical Properties and Improved Hygromechanical Performance. *ACS Appl. Mater. Interfaces* **2015**, *7*, 4595–4607. [CrossRef]
45. Tang, L.; Weder, C. Cellulose Whisker/Epoxy Resin Nanocomposites. *ACS Appl. Mater. Interfaces* **2010**, *2*, 1073–1080. [CrossRef]

nanomaterials

Article

High-Strength Regenerated Cellulose Fiber Reinforced with Cellulose Nanofibril and Nanosilica

Yu Xue [1,†], Letian Qi [2,†], Zhaoyun Lin [2], Guihua Yang [1,2,*], Ming He [2,*] and Jiachuan Chen [2]

1   School of Chemistry and Chemical Engineering, University of Jinan, Jinan 250022, China; xxyy0707@163.com
2   State Key Laboratory of Biobased Material and Green Papermaking, Qilu University of Technology,
    Shandong Academy of Sciences, Jinan 250353, China; lqi01@qlu.edu.cn (L.Q.); linzhaoyun@qlu.edu.cn (Z.L.);
    chenjc@qlu.edu.cn (J.C.)
*   Correspondence: ygh@qlu.edu.cn (G.Y.); heming8916@qlu.edu.cn (M.H.); Tel.: +86-531-8963-1884 (G.Y.);
    +86-531-8963-1861 (M.H.)
†   These authors have contributed equally to this work.

Abstract: In this study, a novel type of high-strength regenerated cellulose composite fiber reinforced with cellulose nanofibrils (CNFs) and nanosilica (nano-SiO_2) was prepared. Adding 1% CNF and 1% nano-SiO_2 to pulp/AMIMCl improved the tensile strength of the composite cellulose by 47.46%. The surface of the regenerated fiber exhibited a scaly structure with pores, which could be reduced by adding CNF and nano-SiO_2, resulting in the enhancement of physical strength of regenerated fibers. The cellulose/AMIMCl mixture with or without the addition of nanomaterials performed as shear thinning fluids, also known as "pseudoplastic" fluids. Increasing the temperature lowered the viscosity. The yield stress and viscosity sequences were as follows: RCF-CNF^2 > RCF-CNF^2-SiO$_2$^{2} > RCF-SiO$_2$^{2} > RCF > RCF-CNF^1-SiO$_2$^{1}. Under the same oscillation frequency, G′ and G″ decreased with the increase of temperature, which indicated a reduction in viscoelasticity. A preferred cellulose/AMIMCl mixture was obtained with the addition of 1% CNF and 1% nano-SiO_2, by which the viscosity and shear stress of the adhesive were significantly reduced at 80 °C.

Keywords: dissolving pulp; cellulose nanofibril; nano silicon dioxide; high strength

Citation: Xue, Y.; Qi, L.; Lin, Z.; Yang, G.; He, M.; Chen, J. High-Strength Regenerated Cellulose Fiber Reinforced with Cellulose Nanofibril and Nanosilica. *Nanomaterials* **2021**, *11*, 2664. https://doi.org/10.3390/nano11102664

Academic Editor: Carla Vilela and Takuya Kitaoka

Received: 2 September 2021
Accepted: 7 October 2021
Published: 11 October 2021

Publisher's Note: MDPI stays neutral with regard to jurisdictional claims in published maps and institutional affiliations.

1. Introduction

Cellulose is a natural and abundant macromolecule that forms with D-glucopyranose ring units in the $^4C_1$-chair configuration, which is linked by β-1, 4-glycosidic bonds that result in an alternate turning of the cellulose chain axis by 180° [1]. Complex hydrogen-bond (H-bond) networks are identified in cellulose macromolecules due to the presence of free hydroxy groups. Thereby, as the complex structure of cellulose makes it difficult to dissolve in water and common organic solvents, there are many restrictions on its use, as well as on its further development [2,3]. Despite the stiff and insoluble nature, the cellulose-based materials have many merits, including being lightweight [4], air permeable [5], etc. As being obtained from plants, they are commonly of low cost, naturally degradable and sustainable [1,6]. Therefore, the dissolution of cellulose is a key step in the production of cellulose-based materials [6]. In the early industrial era, copper ammonia and carbon disulfide methods were commonly used to produce cellulose fiber and film [7]. N-methylmorpholine (NMMO) is currently used in the production of Tencel [8,9]. However, these methods are suffering with severe chemical modification of cellulose, large solvent lost and environmental problems.

The discovery and rapid development of ionic liquids (ILs) reveals their unique solubility and H-bonding ability. Thereby, the application of ILs in dissolving cellulose have been extensively studied [10–13], in order to break the complex H-bond network of cellulose. Swatloski et al. first reported that room-temperature ILs could be used for soluble, non-derived cellulose [14]. Zhang et al. found that cellulose was completely soluble in 1-allyl-3-methylimidazolium chloride (AMIMCl) and 1-butyl-3-methylimidazolium chloride (BMIMCl), which could be used for producing cellulose film and fiber [7,15,16]. ILs such as 1-ethyl-3-methylimidazolium acetate (EMIMAc) have also been reported to prepare cellulose fibers and thin films [17,18]. The dissolved cellulose solution can be spun into regenerated cellulose fibers. Its tensile strength generally lies within the range of 1.6~5.34 cN/dtex, which is comparable to Lyocell [19]. In contrast, the present physical strength of ILs-regenerated fibers failed to meet the standard of high-strength materials, such as industrial textile, rope, triangle belt and so on.

Addition of nanomaterials, even with a small dosage, have been reported to significantly improve the physical strength of the composite materials [20]. Nanocellulose added to composite cellulose paper has been reported to significantly improve its mechanical strength [21]. Nanosilica particles are thoroughly and uniformly dispersed into resin materials, comprehensively enhancing their smoothness and aging resistance [22–24]. Song et al. found microcrystalline cellulose and nano-SiO_2 composite fibers had good tensile strength and thermal stability [23]. Lee et al. used a wet-laid sheet-forming process to produce nanocomposites named CNF/PA6, improving the tensile strengths to 16.4 MPa [25]. He et al. added SiO_2 nanoparticles into cellulose hydrogels to improve the transmittance and mechanical strength [26]. However, most of the literature research studies mainly focus on adding nanomaterials into polymer film and gel system [27,28]. There are only limited studies on the preparation of regenerated cellulose fibers by mixing nanomaterials and pulp in ionic liquid system. Herein, in this work, nanocellulose and nanosilica were added either individually or in combination into softwood-dissolving pulp/1-allyl-3-methylimidazolium chloride (AMIMCl) to improve the physical strength of regenerated fibers. The physical strength of regenerated cellulose fibers was examined, together with chemical, morphological and rheological characterization. The results can provide theoretical support for the preparation of high-strength cellulose fiber.

2. Materials and Methods

2.1. Materials

The bleached softwood dissolving pulp (DP~548; the content of alpha-cellulose was 88.74%.) was obtained from a pulp mill in Jining, China. Nanosilica (BET300 $\pm$ 50 m^2/g, MW 60.08 g/mol, particle size 15 nm) was purchased from Macklin Inc., Shanghai, China. Cellulose nanofibril was obtained from cotton with a diameter of 4–10 nm, length of 1–3 µm, and carboxylic group content of 2.5 mmoL/g, which was purchased from Guilin Qihong Technology Co., Ltd. The 1-allyl-3-methylimidazolium chloride (AMIMCl) IL used in this study was synthesized according to the procedure described in the literature [7]. A successful synthesis was confirmed by nuclear magnetic resonance spectroscopy (AVANCEII400, Bruke Inc., Berlin, Germany), and the water content (0.51%) was verified by a Karl Fischer moisture meter(AKF-1, Mettler Toledo International Co., LTD, Zurich, Switzerland). Methyl imidazole and 3-allyl chloride were purchased from Macklin Inc., Shanghai, China. All of the other chemicals were analytical grade and used without further purification.

2.2. Dissolution and Regeneration Processing of Cellulose

The dissolution process consisted of adding cellulose nanofibers and nanosilica particles into the IL and heating to 70 °C. The dosages of CNF and nano silicon dioxide are shown in Table 1. Dissolved pulp (10 wt% in AMIMCl) was then added into the IL, until the pulp dissolved completely, to generate the cellulose viscose. Then we put the cellulose viscose into a single screw extruder with a single-hole extruder head, and the aperture of extruder head was 0.8 mm. The length–diameter ratio of extruder screw is 23:1. The speed is 50 rad per minute, the amount of viscose extrusion is 300 g/h. All of the fibers used the dry-jet wet-spinning process with a 1.5 cm air-gap and deionized water as a coagulation bath. The fiber was first soaked in deionized water for 24 h, washed with deionized water at 80 °C for 5 min, and then washed with deionized water to remove AMIMCl until the content of AMIMCl in the regenerated fiber was no more than 0.3 wt%. Then, the fibers were dried with hot air at 105 °C for 30 min. Finally, the regenerated cellulose fibers were placed under 23 °C and 50% relative humidity (RH) for 24 h. The fiber without CNF and nano-SiO_2 was named RCF. The fiber with 2% CNF added was named $RCF\text{-}CNF^2$. The fiber with 2% nano-SiO_2 added was named $RCF\text{-}SiO_2{}^2$. The fiber with 1% CNF and 2% nano-SiO_2 added was named $RCF\text{-}CNF^1\text{-}SiO_2{}^1$. The fiber with 2% CNF and 2% nano-SiO_2 added was named $RCF\text{-}CNF^2\text{-}SiO_2{}^2$.

Table 1. Dosage of CNF and nano-SiO_2 in preparation of regenerated cellulose fibers.

Sample	CNF (wt% to Dissolving Pulp)	Nano-SiO_2 (wt% to Dissolving Pulp)
RCF	/	/
$RCF\text{-}CNF^2$	2%	/
$RCF\text{-}SiO_2{}^2$	/	2%
$RCF\text{-}CNF^1\text{-}SiO_2{}^1$	1%	1%
$RCF\text{-}CNF^2\text{-}SiO_2{}^2$	2%	2%

2.3. Characterization of Regenerated Fibers

2.3.1. Tensile Properties of Fibers

The mechanical properties of the regenerated cellulose fibers were analyzed by using a texture analyzer (stable microsystems PL/CEL5, Stable Micro Systems Inc., London, UK), at 23 °C and 50% RH, with a load cell of 5000 g. The initial tensile distance was kept at a constant 20 mm, and the rate of the tensile test was constant at 10 mm/min. The diameter of each fiber sample was measured by using a Vernier caliper for 10 different places per 50 mm. The average diameter was used in the calculations. Each sample was tested at least 10 times.

2.3.2. Contact Angle Test

Since the contact angle of the fiber surface was difficult to measure, it was performed by using cellulose films formed with the viscose extruded from the spinneret head. A water contact angle test was taken on an OCA50 (Dataphysics, Filderstadt, Germany), using 1 μL of liquid.

2.3.3. X-Ray Diffraction (XRD) Analysis

The XRD data were collected on a model D8 ADVANCE diffractometer (Bruker AXS Co., Karlsruhe, Germany) in an angular region 2θ ranging between 5 and 60°, with a scanning speed of 5°/min. The fibers were tightly wound 20 times on the test sheet, covering the test sheet surface completely, and then put into the instrument for testing. Crystallinity was calculated through the diffraction patterns. The calculation method was as follows:

$$\text{Crystallinity } Xc = (I_{101} - I_{am})/I_{101} \times 100\% \tag{1}$$

where I_{101} was the (101) surface diffraction intensity, and I_{am} was the diffraction intensity of the amorphous region. For cellulose I, I_{am} was the diffraction intensity of 2θ at 18.0°. For cellulose II, I_{am} was the diffraction intensity of 2θ at 15.0° [29].

2.3.4. Fourier-Transform Infrared (FTIR) Spectroscopy

The FTIR spectra of the regenerated cellulose fibers were recorded by using an ALPHA FTIR spectrophotometer with an attenuated total reflectance (ATR) (Bruker Corporation, Billerica, MA, USA). All of the spectra were obtained from 32 scans with a resolution of $4\ cm^{-1}$ and absorption mode, using a wavelength range from 500 to $4000\ cm^{-1}$. At least three repetitions per sample were conducted.

2.3.5. Scanning Electron microscopy (SEM) and Energy-Dispersive X-ray Spectroscopy (EDS)

Images of the regeneration cellulose fiber samples were taken on a regulus8220 (Hitachi Ltd., Tokyo, Japan) microscope operating at 5 kV. The fracture surface of the fibers was coated with a platinum layer (with a thickness of approximately 20 nm) before observation. Moreover, the surface element content of the fibers was obtained via EDS on a regulus8220 (Hitachi Ltd., Tokyo, Japan) at 10 kV.

2.3.6. Thermal Analysis

Thermographic analysis (TGA) of the fibers was performed on a simultaneous thermal analyzer (TA TGAQ50, TA Instruments, New Castle, DE, USA). A total of 30~40 mg of the sample was placed in an aluminum oxide pan and heated from 35 to 800 °C, at a rate of 10 °C/min, in a nitrogen atmosphere, with a flush rate of 40 mL/min.

2.3.7. Rheological Measurements

Steady and dynamic rheology experiments were carried out on an TA ARES-G2 rotated rheometer (TA Instruments, New Castle, DE, USA)). Parallel plates (20 mm in diameter) were used. The chosen gap was 1 mm for all of the measurements. The strain amplitude values were checked to ensure that the measurement deviation did not exceed 5%, which was in the linear viscoelastic regime. The shear rate ranged from 0.1 to 100 s^{-1}, and the sweep of the angular frequency (w) was 6.28 rad/s. Dynamic viscoelastic functions, such as the shear storage modulus (elasticity modulus) (G′) and loss modulus (viscous modulus) (G″) as a function of time, angular frequency, and temperature, were also measured.

3. Results and Discussion

3.1. Tensile Properties

It was reported that the addition of nanomaterials could increase the strength of polymers [24,30], similar results were found in this work. As can be seen from Table 2, the tensile strength of RCF is 155.82 MPa, and the elongation at break is 8.73%. After adding CNF and nano-SiO_2, the tensile strength of all the regenerated cellulose fibers was improved, as the tensile strength of RCF-CNF[2] is 174.15 MPa, which is increased by 11.76% in comparison with RCF. Its broken number in spinning is 5, and the elongation at break is 7.43%, which is decreased by 14.89%. Similarly, the tensile strength of RCF-SiO_2[2] is 174.18 MPa, which is increased by 11.76% than that of RCF. Its elongation at break is 7.65%, which is decreased by 12.37%. Therefore, the tensile strength of RCF-CNF[2] and RCF-SiO_2[2] increase by the same extent.

Table 2. Tensile properties of the regenerated cellulose fibers prepared with or without nanomaterials.

Sample	Tensile Strength (MPa)	Elongation at Break (%)	Number of Broken in Spinning (/h)
RCF	155.82 ± 5.23	8.73 ± 0.13	1
RCF-CNF[2]	174.15 ± 2.36	7.43 ± 0.11	5
RCF-SiO_2[2]	174.18 ± 1.26	7.65 ± 0.15	3
RCF-CNF[1]-SiO_2[1]	229.78 ± 0.91	10.01 ± 0.15	2
RCF-CNF[2]-SiO_2[2]	173.42 ± 3.35	6.33 ± 0.09	2

The enhancement of physical strength implied that the nanomaterials added closely interacted with regenerated cellulose, instead of just attaching on the fiber surface. Both nanocellulose and softwood-dissolving pulp can be dissolved in AMIMCl because of the strong coulombic and H-bond interactions within the IL system [31]. Interacted by the IL anions and cations, the intermolecular and intramolecular hydrogen bonds of cellulose chains are broken, which results in the cellulose dissolution and formation of homogeneous solution [15,32]. The length of softwood dissolving pulp is normally 0.5~0.6 mm, whereas the typical length of CNF is 0.001~0.003 mm. Since CNF has a short molecular chain, large specific surface area and free active carboxyl groups, during the regeneration process, it can interact closely with cellulose fiber chains through H-bonding, resulting in the enhancement of their physical strength.

In terms of nano-SiO_2, as an inorganic compound, it was added into cellulose viscose system to form an organic–inorganic hybrid viscose system. As nano-SiO_2 is difficult to dissolve in AMIMCl, it was only dispersed as suspensions. The excellent specific surface area allowed nano-SiO_2 to have strong interfacial adsorption through van der Waals interactions with cellulose [23]. Nano-SiO_2 can also bind with the small amount of water presented in cellulose/ILs system, where nano-SiO_2, cellulose, and water are tightly bonded to form a complex network structure [26,33].

According to the literature, it was expected that increasing dosage of nanomaterials would lead to a further enhancement in strength [23]. However, the fiber strength of RCF-CNF^2-$SiO_2{}^2$ was the same as RCF-CNF^2 and RCF-$SiO_2{}^2$. The tensile strength of RCF-CNF^2-$SiO_2{}^2$ was 173.42 MPa, which was increased by 11.29% than that of RCF. Its elongation at break of RCF-CNF^2-$SiO_2{}^2$ was 6.33%, which was decreased by 27.49%. Interestingly, it was found that the addition of 1% CNF and 1% nano-SiO_2 resulted in a significant increase in the fiber strength, and the elongation at break also dramatically increased. It should be noticed that, in all other testing samples, the elongation at break were all decreased compared with the RCF sample. The tensile strength of RCF-CNF^1-$SiO_2{}^1$ was 229.78 MPa, an increase of 47.46%, which was the strongest of all the fibers; and the elongation at break of RCF-CNF^1-$SiO_2{}^1$ was 10.01%, an increase of 14.66%. This indicated that RCF-CNF^1-$SiO_2{}^1$ may make the viscose system and regeneration process different from other conditions, in order to provide a tightly interacted composite fiber.

3.2. Water Contact Angle

Nano-SiO_2 is a common surface hydrophobic agent that can give fibers a particular hydrophobic property when coated on their surface [24]. Cellulose has excellent hydrophilicity due to the presence of a large number of hydroxyl groups in its structure [33–35]. Therefore, the addition of these nanomaterials should alter the hydrophilicity of the regenerated fibers. Since the contact angle of the fiber surface was difficult to measure, it was performed by using cellulose films formed with the viscose extruded from the spinneret head.

Results in Figure 1 show the change of contact angle with the addition of nanomaterials. The cellulose film without nanomaterials was 51.85°, whereas the contact angles of the cellulose film with RCF-CNF^2, RCF-$SiO_2{}^2$, RCF-CNF^1-$SiO_2{}^1$, and RCF-CNF^2-$SiO_2{}^2$ were 62.1, 69.15, 67.05, and 78.9°, respectively. Therefore, the hydrophily of all the films with nanomaterials added was reduced. It was assumed that the addition of nanocellulose interacted with regenerated cellulose, making the surface smooth and dense, and thus resulting in greater contact. Meanwhile, as a hydrophobic agent, the presence of nano-SiO_2 on the surface reduced hydrophilicity and increased the contact angle [33].

3.3. XRD Analysis

Figure 2 and Table 3 showed the XRD patterns and crystallinity results of raw materials and fibers with or without nanomaterials added. As can be seen from the XRD images, the configuration of conifer pulp was of cellulose I type crystallographic form, and the regenerated cellulose fibers made were all of cellulose II form. The bleached softwood dissolving pulp was dissolved in AMIMCl, and the hydrogen bonds between the cellulose molecules were broken. The cellulose solution entered the coagulant through the extruder head, after which the AMIMCl in the viscose dissolved in the water, and hydrogen bonds between the cellulose molecules reformed. Glucose units were bonded together by a secondary helix structure during the formation process to produce cellulose II type of crystallography. Meanwhile, the results showed that the addition of CNF and nano-SiO_2 did not affect the crystal structure of cellulose in the regeneration process, which also illustrated that the CNF added was also dissolved in AMIMCl and regenerated into cellulose II [1,15,36].

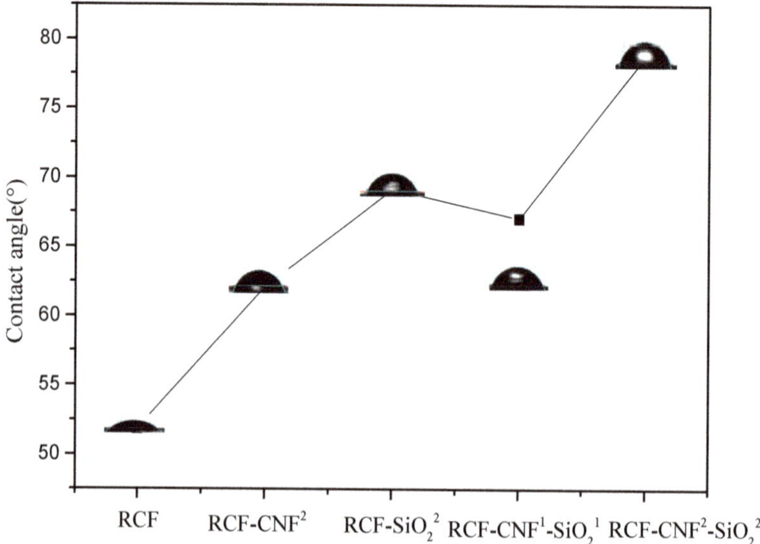

Figure 1. Contact angle of the regenerated cellulose films prepared with or without nanomaterials.

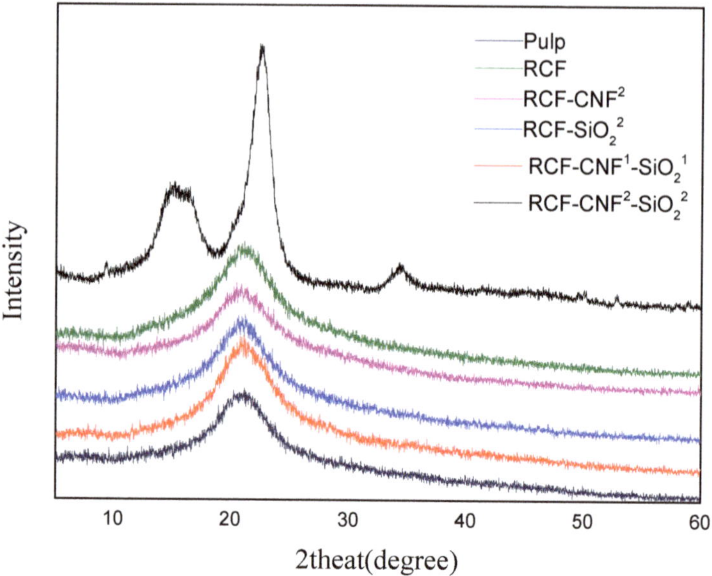

Figure 2. XRD analysis of the regenerated fibers with or without nanomaterials versus control cellulose pulp.

Table 3. Crystallinity of the regenerated fibers with or without nanomaterials versus control cellulose pulp.

Sample	Crystallinity (%)
Pulp	$50.88 \pm 1.03\%$
RCF	$62.76 \pm 1.21\%$
RCF-CNF^2	$50.11 \pm 0.96\%$
RCF-SiO$_2$^{2}	$47.37 \pm 1.54\%$
RCF-CNF$^{1\text{-}}$SiO$_2$^{1}	$55.15 \pm 0.53\%$
RCF-CNF$^{2\text{-}}$SiO$_2$^{2}	$47.75 \pm 0.75\%$

During the spinning process, the cellulose macromolecular chains were oriented by high shear and subsequent stretching in the air gap, which induced crystallization during drying. It is known that the tenacity of regenerated fibers mainly depends on the orientation of their amorphous region, while the modulus and crystallinity are related to crystal orientation [19]. Thus, the crystallinity and orientation determine the mechanical properties of the regenerated fibers.

The crystallinity of the pulp was 50.88%, and the crystallinity of RCF was 62.76%, with an increase of 10%. This was because the hydrogen bond length of cellulose II was shorter than that of cellulose I. As a result, the cellulose molecular chains were more tightly stacked, and their structure was more thermodynamically stable [37]. The crystallinity of RCF-CNF^2 was 50.11%, which was about 20.15% lower than that of RCF. The crystallinity of RCF-SiO$_2$^{2} was 47.37%, which was the lowest among the samples. The crystallinity of RCF-CNF^2-SiO$_2$^{2} was 47.75%, which was basically the same as that of RCF-SiO$_2$^{2}. However, the crystallinity of RCF-CNF^1-SiO$_2$^{1} sample was significantly elevated to 55.15%, which was higher than those of other nanomaterials added samples. However, this value was still lower than that of RCF sample. Therefore, it could be concluded that the crystallinity of the fibers decreased with the addition of CNF and nano-SiO$_2$.

It is generally believed that fibers with higher crystallinity usually present stronger physical strength, which could be attributed to the rigid and stiff nature of crystalline regions [19,23,24]. However, the results presented in this work, where samples with the addition of nanomaterials showed lower crystallinity but higher tensile strength. Therefore, the characterization over their chemical interactions, morphology and rheological behaver would be crucial to understanding the mechanism of strength enhancement.

3.4. Fourier-Transform Infrared (FTIR) Spectroscopy Analysis

Considering the fact that all testing samples contained ~98% of cellulose, the FTIR spectra in Figure 3 shows similar trend, indicating that the structure and composition of the fibers were similar. The peak in the range 2900~2800 cm^{-1} was related to stretching of C–H from methyl groups. The peak in the range 1400~1150 cm^{-1} corresponded to the deformations of C=OH. The peak in the range 1135~1180 cm^{-1} corresponded to the deformations of the C–OH, C–H, and C–O–C groups of cellulose. The bond at 1100 cm^{-1} was considered to be the characteristic peak of C–C. The 887 cm^{-1} peak was characteristic of the β-glycoside bond between the glucose units. The peak in the range 700~450 cm^{-1} corresponded to the vibration bending of O–H. The 804 cm^{-1} peak corresponded to the symmetric stretching of Si–O–Si [23,24,38–42], indicating the presence of SiO$_2$ on the fiber surface.

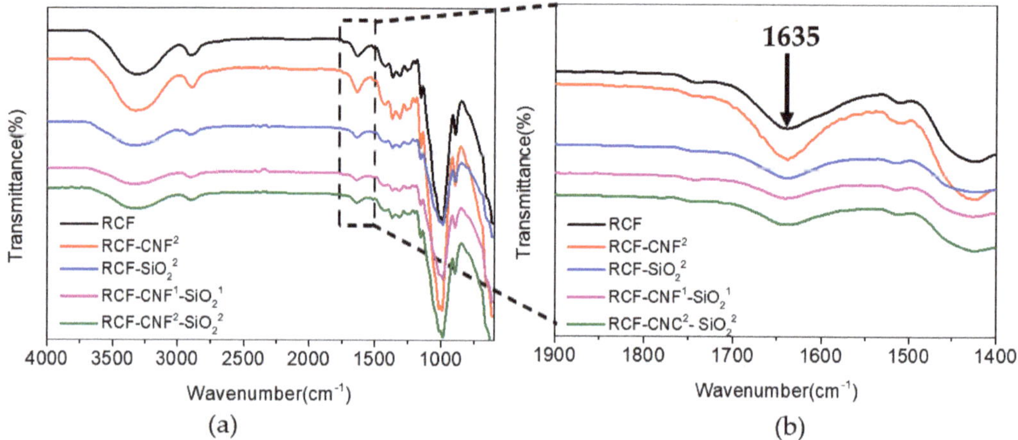

Figure 3. FTIR spectra of RCF, RCF-CNF^2, RCF-SiO$_2$^{2}, RCF-CNF$_1$-SiO$_2$^{1}, and RCF-CNF^2-SiO$_2$^{2}: (**a**) Transmittance of fibers between the wavelength range from 4000 to 500 cm^{-1}; (**b**) transmittance of fibers between the wavelength range from 1900 to 1400 cm^{-1}.

However, minor difference in the intensity of the peaks could be identified. The band in the range 3600~3200 cm^{-1} is corresponded to the vibrational modes of the hydroxyl group in the cellulose molecules, particularly the band at 3500 cm^{-1} represented the stretching characteristic of OH groups, and the bond at 3246 cm^{-1} was the characteristic peak of hydrogen bonding. The changes in the OH groups' related peaks indicate that the addition of CNF causes an increase in free OH groups, whereas the addition of nano-SiO$_2$ results in a significant reduction of them. Similar trend was noticed at 1640~1630 cm^{-1}, which was related to the water adsorption of the fibers. In comparison with the RCF sample, water adsorption in RCF-CNF^2 increased, while RCF-SiO$_2$^{2} was significantly reduced. This result may attribute to the hydrophobicity of the silica nanoparticles. However, it is interesting to note that the addition of CNF at the same dosage cannot overcome the reduction caused by nano-SiO$_2$, where RCF-CNF^1-SiO$_2$^{1} and RCF-CNF^2-SiO$_2$^{2} samples present similar bands with that of RCF-CNF^2.

Therefore, the infrared spectrum analysis shows that no derivatization reaction appears during cellulose regeneration. It has been reported that the cellulose dissolution and regeneration in AMIMCl is a physical process [7], which mainly involves the breaking and recombination of hydrogen bonds without chemical reactions. Thus, it is believed that the increase of fiber strength in this work was mainly caused by the rheology properties of viscose and the action of secondary bonds, including hydrogen bonding and interface adsorption [24].

3.5. SEM and EDS Analysis

The SEM images shows the surface morphology of the raw materials (Figure 4(a1,a2)) and regenerated fibers. As expected, the softwood dissolving pulp was a flat fiber with a layered fiber structure on its surface (Figure 4a) and some fine and filamentation fibers on the surface (Figure 4(a2)). After regeneration via dry spraying and wet spinning, the fiber surface presented as a smooth, filamentous structure (Figure 4b). Moreover, the fiber surface was a dense, scaly structure with pores and micro protrusions (Figure 4(b2)). The fiber surface with RCF-CNF^2 added was relatively smooth (Figure 4(c1,c2)). The RCF-SiO$_2$^{2} had some protuberance prominences on the surface (Figure 4(d1,d2)). The appearance of small particles may attribute to the aggregation of nano-SiO$_2$ nanoparticles and the incompatibility of organic and inorganic materials. Meanwhile RCF-CNF^1-SiO$_2$^{1} had the densest and smoothest surface structure (Figure 4(e1,e2)), and the RCF-CNF^2-SiO$_2$^{2} was also relatively dense but with more prominences (Figure 4(f1,f2)). Adding CNF and nano-SiO$_2$ simultaneously makes the material smooth and flat. The literature has reported similar results, wherein the addition of CNF and nano-SiO$_2$ resulted in a smooth composite surface and could lead to an enhancement of mechanical properties [22,23,43].

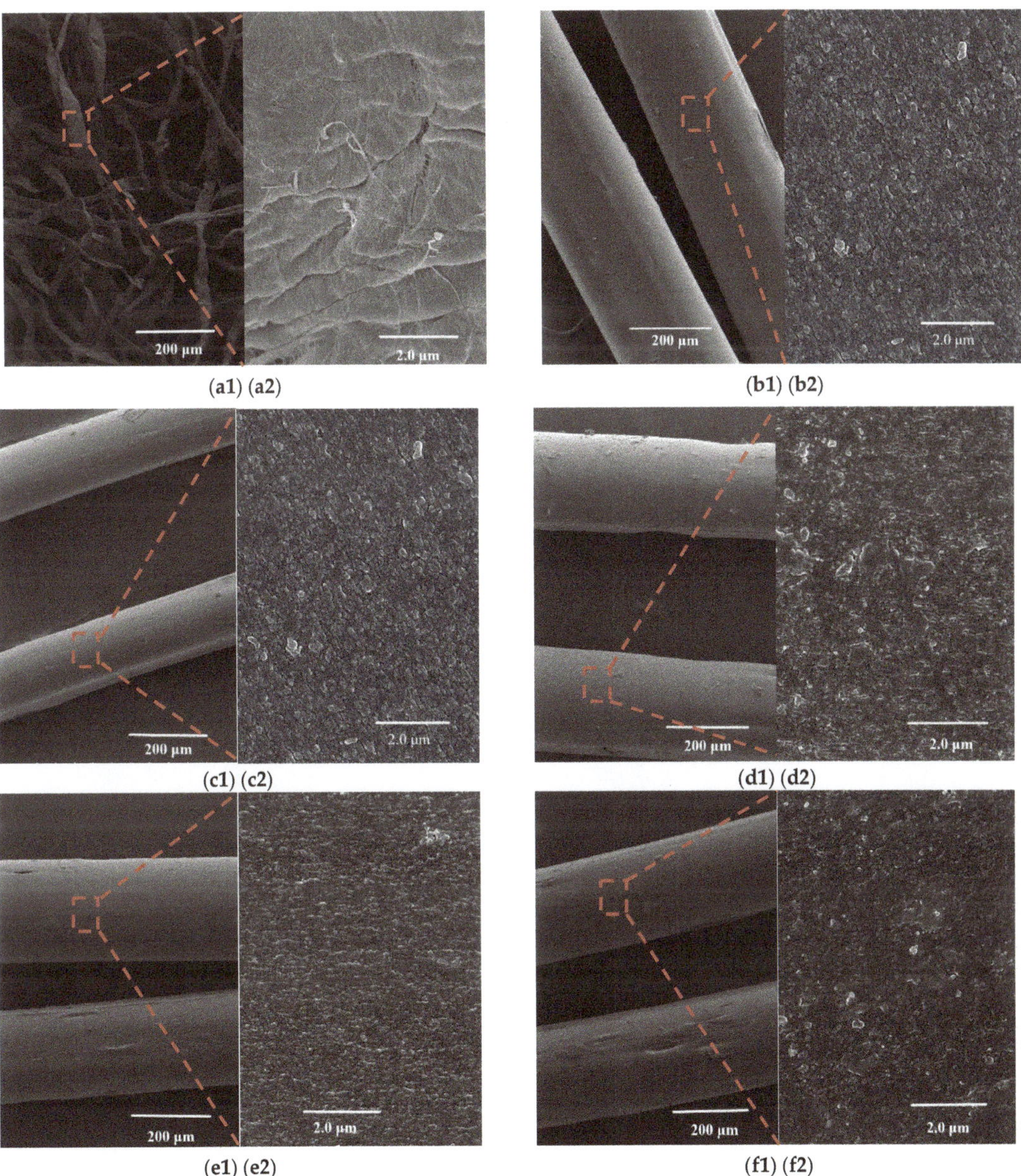

Figure 4. *Cont.*

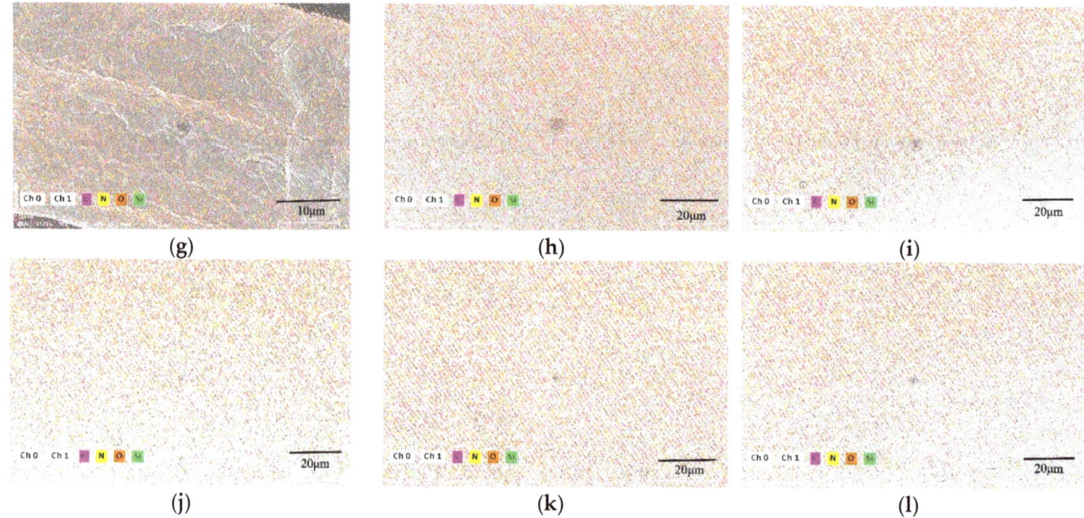

Figure 4. SEM and EDS images of the regenerated cellulose fibers: SEM images of pulp at ×100 (**a1**) and ×10,000 (**a2**), RCF at ×100 (**b1**) and ×10,000 (**b2**), RCF-CNF^2 at ×100 (**c1**) and ×10,000 (**c2**), RCF-SiO$_2{}^2$ at ×100 (**d1**) and ×10,000 (**d2**), RCF-CNF^1-SiO$_2{}^1$ at ×100 (**e1**) and ×10,000 (**e2**), RCF-CNF^2-SiO$_2{}^2$ at ×100 (**f1**) and ×10,000 (**f2**); EDS images of Pulp at ×2000 (**g**), RCF at ×1000 (**h**), RCF-CNF^2 at ×1000 (**i**), RCF-SiO$_2{}^2$ at ×1000 (**j**), RCF-CNF^1-SiO$_2{}^1$ at ×1000 (**k**), and RCF-CNF^2-SiO$_2{}^2$ at ×1000 (**l**).

It can be seen via EDS analysis that CNF and nano-SiO$_2$ were retained in the fiber (Table 4), as the Si content of RCF and RCF-CNF^2 was 0.15%, while the Si content of RCF-SiO$_2{}^2$, RCF-CNF^1-SiO$_2{}^1$, and RCF-CNF^2-SiO$_2{}^2$ was 0.56%, 0.49%, and 0.70%, respectively. The distributions of Si on the fibers' surface were uniform (Figure 4g–l). However, the silica content on the surface was lower than the added amount. Therefore, it was speculated that most of the silicon atoms were embedded inside the cellulose fibers, where strong interactions between nano-SiO$_2$ and cellulose macromolecules formed [23,44,45].

Table 4. Relative element content of the regenerated fibers surface with or without nanomaterials and control cellulose pulp by EDS.

Sample	C (%)	O (%)	N (%)	Si (%)
Pulp	28.2	46.32	25.35	0.14
RCF	28.05	46.63	25.16	0.15
RCF-CNF^2	27.48	47.51	24.85	0.15
RCF-SiO$_2{}^2$	27.42	47.35	24.67	0.56
RCF-CNF^1-SiO$_2{}^1$	26.87	48.53	24.11	0.49
RCF-CNF^2-SiO$_2{}^2$	27.51	47.23	24.56	0.70

3.6. Thermal Analysis

Figure 5 shows that the regenerated cellulose fibers all exhibited typical features of cellulose degradation. The first degradation stage was from room temperature to about 150 °C and was mainly associated with the fiber dehydration and volatilization of some unstable substances. The second stage involved the decomposition of the amorphous region in the fiber structure, and part of the glucose group began to dehydrate. The temperature of this stage was generally between 150 and 250 °C. The fastest degradation rate occurred between 250 and 400 °C, during which the glycosidic bonds of cellulose broke and cellulose began to degrade [46–48].

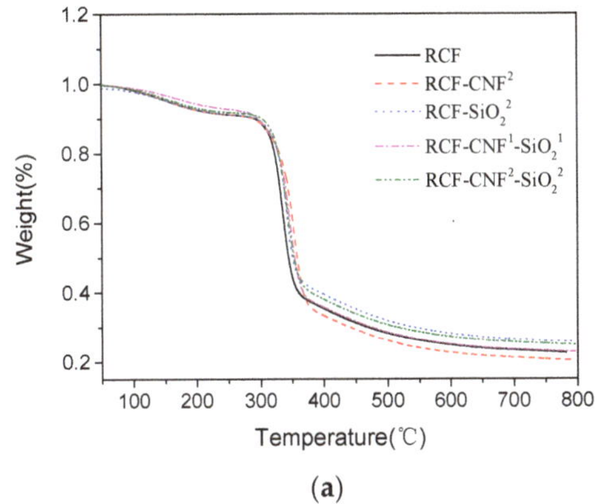 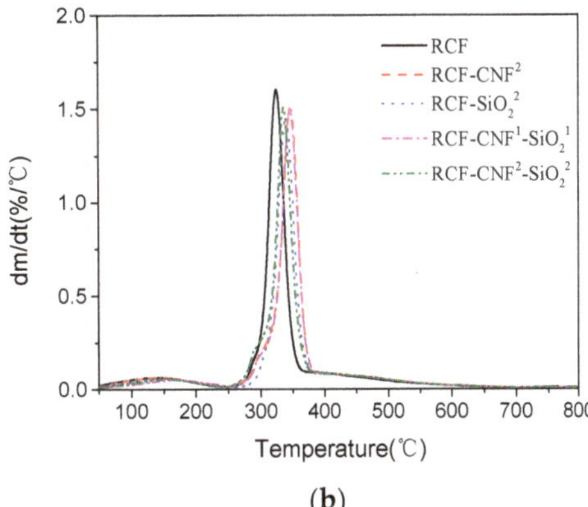

Figure 5. TGA (**a**) and DTA (**b**) curve of the regenerated cellulose fibers with or without nanomaterials.

The rapid degradation temperature of RCF-CNF^2 was lower than the other four samples because the nanostructure of CNF made it easy to degrade, reducing the temperature of large-scale degradation of the fibers. The fastest degradation rate of RCF-SiO$_2$^{2} was lower than RCF, and the thermal stability of the cellulose was slightly improved with the addition of 2% nano-SiO$_2$. The distribution of silica can prevent the migration of small molecules in the fibers, so the decomposition of the cellulose macromolecular chain was delayed in the process of thermal decomposition [23,24].

From Table 5. it can be seen that final residual amount of fiber without nanomaterial was 22.58%. RCF-CNF^2 started to degrade earlier than other samples, and the final residual amount was 20.39%. The fiber with added 2% CNF was completely degraded. The residual amount of the fiber with added 2% nano-SiO$_2$ was 25.80%. The residual amount of RCF-CNF^1-SiO$_2$^{1} was 23.05%, and the residual amount of RCF-CNF^2-SiO$_2$^{2} was 24.93%. These results indicating that only minor loss of nano-SiO$_2$ occurred in the regeneration process [23]. By taking the EDS results into consideration, this confirmed the previous hypothesis that most of the nano-SiO$_2$ was embedded into the cellulose fibers during the regeneration process.

Table 5. Thermal properties of the regenerated cellulose fibers after different stages of treatment.

Sample	T_{on} (°C)	Mass Loss (%)	T_{max} (°C)	Mass Loss (%)	Residue at 800 °C (%)
RCF	307	12.28	329	27.16	22.48
RCF-CNF^2	300	10.96	344	30.44	20.39
RCF-SiO$_2$^{2}	308	11.76	343	30.42	25.80
RCF-CNF^1-SiO$_2$^{1}	302	10.96	344	36.25	22.85
RCF-CNF^2-SiO$_2$^{2}	313	11.76	335	26.42	24.93

3.7. Rheological Properties of Cellulose/IL Solutions with Different Contents of Nanomaterial

In the process of viscose regeneration, the fluidity of viscose has a direct effect on the properties of regenerated cellulose fibers, and the viscosity is closely related to the phenomenon of filament breaking and filament doubling in the spinning process. To explore the reasons for the changes of fiber strength and crystallinity, the rheological properties of the viscose system during the extruding was tested. The viscose sample was obtained at the extruder head. The temperature of viscose at the extruder head was 80 °C and that of coagulation bath was 25 °C. Thus, the viscosity, stress, and viscoelasticity analysis of the viscose was carried out at 40, 60, and 80 °C.

From Figure 6, it can be seen that all testing viscose systems were shear-thinning system. The yield stress and viscosity sequence were as follows: RCF-CNF^2 > RCF-CNF^2-SiO$_2$^{2}

> RCF-SiO$_2^2$ > RCF > RCF-CNF^1-SiO$_2^1$ viscose. For a low shear rate, the shear force produced a slight shearing orientation effect on the viscose, but the components in the system were still in a state of free distribution, and the shear viscosity was basically unchanged.

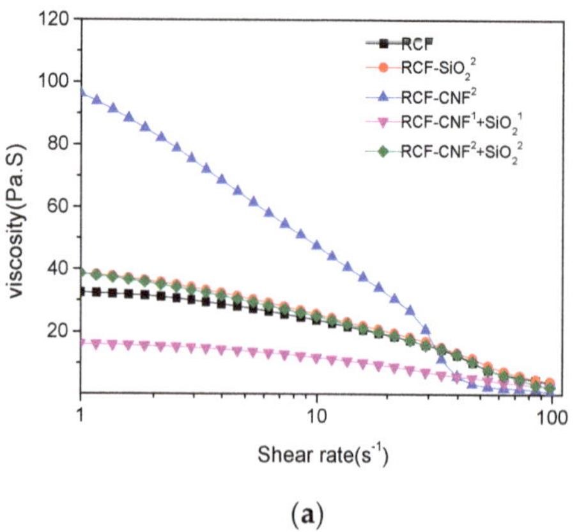
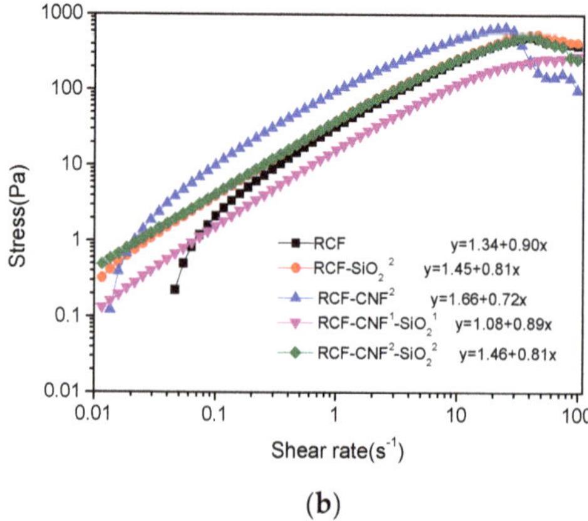

Figure 6. Viscosity (**a**) and shear stress (**b**) of cellulose viscose systems with or without nanomaterials at 80 °C.

RCF-CNF^2 had the largest initial viscosity of all the samples at 80 °C. This was mainly because CNF was identical to the wood fibers in terms of chemical structure, making it evenly dissolve and distribute in the viscose. Thus, the network structure of cellulose and ILs formed, and within the viscose, the CNF, pulp, and IL formed good crosslinking and hydrogen bonds to become a homogeneous mixing system characterized by a high yield stress value [2].

Comparatively, the viscosities of RCF-CNF^2-SiO$_2^2$ and RCF-SiO$_2^2$ were lower than RCF-CNF^2 viscose, but the viscosities of RCF-CNF^2-SiO$_2^2$ and RCF-SiO$_2^2$ and the viscosity trend were of the same. This may be due to the nano-SiO$_2$ reducing the generation of hydrogen bonds among the components [23], which were replaced by interfacial adsorption, thus lowering the overall viscosity. The viscosity of RCF-CNF^1-SiO$_2^1$ was the lowest of all the samples, which meant that this system had less entanglement of macromolecules and force among components, or less network structures, better fluidity of the viscose, and more continuous viscose in the process of extruding. The force between each component was less affected by the pressure change of the spinneret, so it was difficult to break and strand, which was more conducive to the regeneration of cellulose and made the fiber surface smoother. This was consistent with the surface morphology characterized by SEM, which was also the reason for the higher fiber strength.

From the results shown in Figure 6, we can observe that, with the increase of shear rate, the apparent viscosity of several kinds of solutions decreased obviously to show shear thinning, which is typical of non-Newtonian fluids. The entangled macromolecules began to align in a directional manner, showing a decrease in apparent viscosity. When cellulose was completely dissolved in the IL, the van der Waals forces produced physical crosslinking between the molecules [49,50]. Hydrogen bonds existed between the cellulose macromolecules and macromolecules, and between cellulose the macromolecules and ions. The Brownian motion of each component in the viscose system made the physical crosslinking and hydrogen bonds exist in a dynamic equilibrium of constant destruction and regeneration, which made the viscose form a network structure. With the increase of shear force, the dynamic equilibrium was destroyed; the physical crosslinking and hydrogen bonds were not ready to recombine; and there appeared a relative slip and rearrangement between the components, resulting in a decrease in viscosity that eventually reached a constant value [51–53]. The shear rate had little influence on the viscosity of

RCF-CNF^1-SiO$_2$^{1}; that is to say, the state of the whole system was more stable, even during extrusion, and the tensile strength of the fiber was better.

Figure 7 shows the storage modulus G′ and loss modulus G″ of the testing viscose. For the viscose system without nanomaterials, G′ was greater than G″ at 40 °C, indicating that the elastic modulus was greater than the viscous modulus, which showed typical solid-like properties. The main forces of the system were based on the entanglement between cellulose molecules and the network structure between cellulose/AMIMCl. At temperatures of 60 and 80 °C, G″ was greater than G′, indicating that the viscous modulus was greater than the elastic modulus, showing fluid properties, which also indicated that the properties of viscose were sensitive to the temperature. As the temperature changed, the viscoelasticity also changed. With the increase of temperature, the cellulose macromolecules were more active in AMIMCL, making the viscose appear fluid.

The viscoelasticity of viscose mainly depends on the properties of each component and their interactions. It is generally believed that the addition of nano-materials can change the viscoelasticity of viscose. For viscose with added CNF and nano-SiO$_2$, when the temperature was constant, with the increase of frequency, the solution G′ and the solution G″ both increased. Furthermore, values exhibited an intersection point, which is called the fiber solution gel point. With the increase of temperature, the intersection point moved toward the high-frequency area, where modulus transposition occurs, and the sample experienced structure reconstruction and an elasticity larger than the viscosity, which are typical characteristics of an entangled polymer solution. Under the same oscillation frequency, when the temperature increased, G′ and G″ decreased, which indicated that the viscoelasticity decreased [54–56].

At 80 °C, G′ and G″ of RCF-CNF^2 were the largest among all of the samples, which indicated that the entanglement degree between the fiber molecules in the viscose system was higher, and the physical entanglement and hydrogen bonds among the cellulose macromolecules, CNF, and IL formed a tangled network structure that affected the relative motion and stretching of the molecules. Therefore, the viscose showed an obvious elasticity. The intersection point of G′ and G″ experienced a minimal oscillation frequency, indicating that as adding CNF increased the cellulose content within the viscose system, the gel point of the whole system moved forward. Thus, in the regeneration process, solidification was easier. The G′ and G values of RCF-SiO$_2$^{2}, RCF-CNF^1-SiO$_2$^{1}, and RCF-CNF$_2$-SiO$_2$^{2} were all lower than those of the blank samples, indicating that adding nano-SiO$_2$ made the network structure in the system worse and reduced the generation of partial hydrogen bonds.

As shown in Table 6, for RCF-SiO$_2$^{2} at a temperature of 80 °C, the intersection points between G′ and G″ experienced an oscillation frequency of 102.78 rad/s, with a higher angular frequency of gelatinization. For RCF-CNF^2-SiO$_2$^{2} and RCF-CNF^1-SiO$_2$^{1}, their G′ were similar to G″, which indicated that adding equal amounts of CNF and nano-SiO$_2$ in the system basically had the same effect on the entangling of macromolecules and changing the forces between the components. For RCF-CNF^1-SiO$_2$^{1}, its intersection points of G′ and G″ experienced an oscillation frequency of 90.28 rad/s. For RCF-CNF^2-SiO$_2$^{2}, the intersection points of G′ and G″ was at 100.79 rad/s. In the RCF-CNF1-SiO21 system, the degree of aggregation and entanglement among the components was greater than that of RCF-CNF^2-SiO$_2$^{2}.

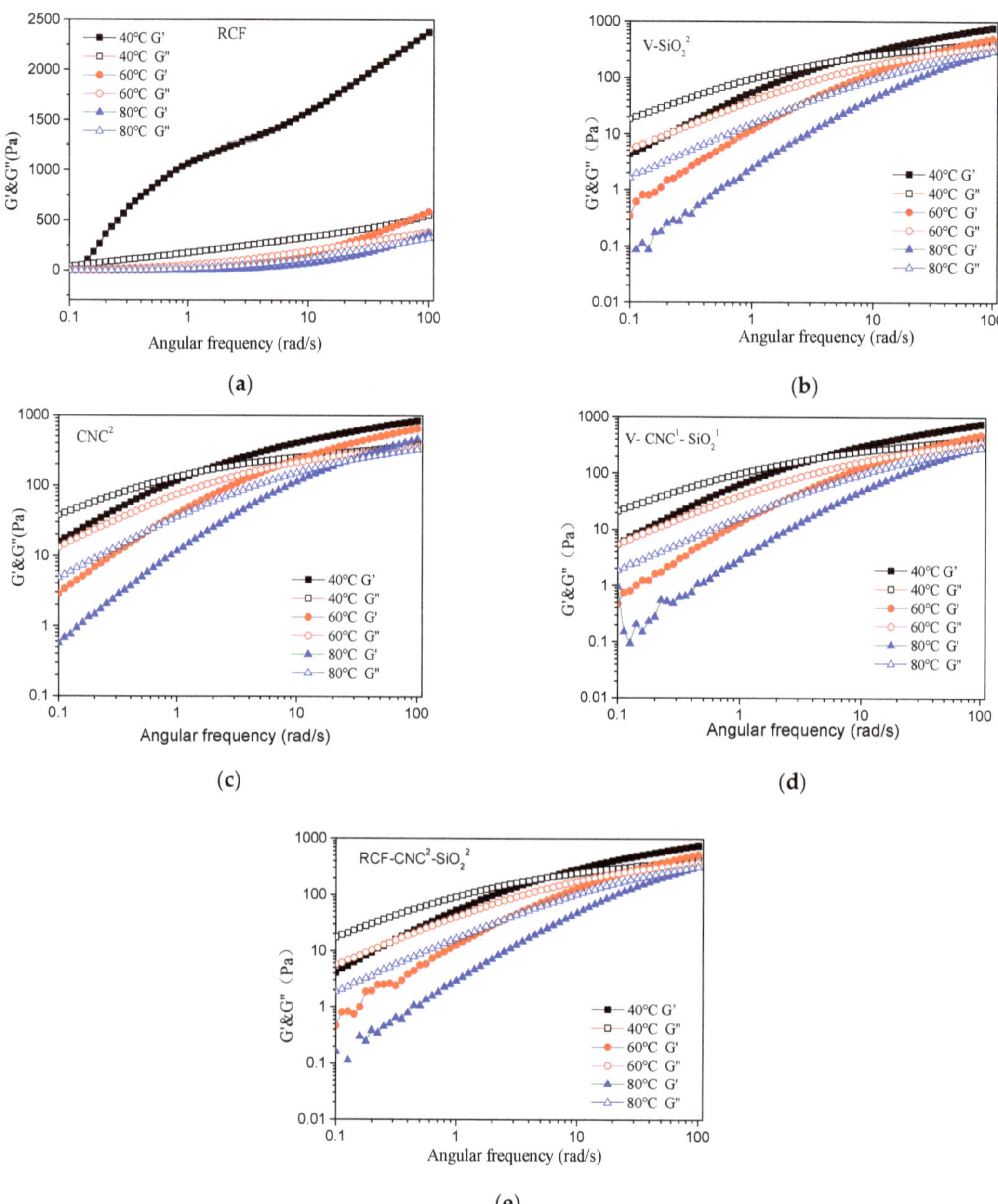

(**a**)

(**b**)

(**c**)

(**d**)

(**e**)

Figure 7. Influence of temperature and angular frequency on G′ and G″ of RCF (**a**), RCF-CNF^2 (**b**), RCF-SiO$_2$^{2} (**c**), RCF-CNF^1-SiO$_2$^{1} (**d**), and RCF-CNF^2-SiO$_2$^{2} (**e**).

Table 6. Modulus values and frequency at the crossover point of G′ and G″ of cellulose viscose.

Sample	Temperature (°C)	Angular Frequency (Rad/s)	Modulus Values (Pa)
RCF	40	0.13	56.45
	60	13.39	220.34
	80	59.46	282.61
RCF-CNF^2	40	1.59	161.68
	60	7.117	195.78
	80	25.21	221.03
RCF-SiO$_2$^{2}	40	5.27	207.32
	60	25.41	237.24
	80	102.78	282.66
RCF-CNF^1-SiO$_2$^{1}	40	4.50	191.22
	60	22.62	221.78
	80	90.28	283.62
RCF-CNF^2-SiO$_2$^{2}	40	5.67	207.325
	60	25.25	251.22
	80	100.79	324.53

4. Conclusions

In this study, conifer pulp was used as a raw material to make regenerated fiber, an ionic liquid AMIMCl was used as a solvent for cellulose, and organic CNF fiber and inorganic nano-SiO$_2$ were added to regenerate composite cellulose fibers via a dry-jet wet-spinning process. The mechanical properties of the composite cellulose can be improved by adding nanomaterials. Adding 1% CNF and 1% nano-SiO$_2$ to the pulp/AMIMCl improved the mechanical properties of the composite cellulose by 47.46%. The regenerated fibers and nanocomposite fibers were typical type-II cellulose crystalline forms, and the crystallinity of the composite fibers decreased. The surface of the regenerated fiber exhibited a scaly structure with pores, and the pores could be reduced by adding 1% CNF and 1% nano-SiO$_2$. The addition of nanomaterials changed the degradation rate of the regenerated fibers. From the perspective of the residual amount, the loss rate of nanomaterials in the preparation process was low. The viscosities of all five samples were pseudoplastic fluids with a yield value of stress and with shear thinning. The yield stress and viscosity sequences were as follows: RCF-CNF^2 > RCF-CNF^2-SiO$_2$^{2} > RCF-SiO$_2$^{2} > RCF > RCF-CNF^1-SiO$_2$^{1}. Under the same oscillation frequency, when the temperature increased, the G′ and G″ decreased, indicating a decrease in viscoelasticity.

Author Contributions: Y.X. and L.Q. contributed equally to this work. L.Q., G.Y. and M.H. conceived and designed the experiments; Y.X., L.Q. and Z.L. performed the experiments; Y.X., L.Q. and Z.L. analyzed the data; G.Y. and M.H. contributed reagents, materials, and analysis tools; Y.X., L.Q. and Z.L. wrote the study; G.Y., M.H. and J.C. reviewed the manuscript and made comments. All authors have read and agreed to the published version of the manuscript.

Funding: The authors are grateful for the financial support received from the National Natural Science Foundation of China (Grant Nos. 31770628, 31901267 and 22108134), the Provincial Key Research and Development Program of Shandong (Grant Nos. 2019JZZY010326 and 2019JZZY010328), Foundation of State Key Laboratory of Biobased Material and Green Papermaking (No. ZZ20190214), and the Pilot Project for Integrating Science, Education and Industry (Grant No. 2020KJC-ZD14).

Data Availability Statement: All date used during the study appear in the submitted article.

Acknowledgments: The authors are grateful for the financial support received from the National Natural Science Foundation of China (Grant Nos. 31770628, 31901267 and22108134), the Provincial Key Research and Development Program of Shandong (Grant Nos. 2019JZZY010326 and 2019JZZY010328), Foundation of State Key Laboratory of Biobased Material and Green Papermaking (No. ZZ20190214), and the Pilot Project for Integrating Science, Education and Industry (Grant No. 2020KJC-ZD14).

Conflicts of Interest: The authors declare that they do not have any conflict of interest.

References

1. Rojas, O.J. Cellulose chemistry and properties: Fibers, nanocelluloses and advanced materials preface. In *Cellulose Chemistry and Properties: Fibers, Nanocelluloses and Advanced Materials*; Rojas, O.J., Ed.; Springer International Publishing: Cham, Switzerland, 2016; Volume 271, pp. V–VI, 3–9.

2. Casado, U.; Mucci, V.; Aranguren, M.I. Cellulose nanocrystals suspensions: Liquid crystal anisotropy, rheology and films iridescence. *Carbohydr. Polym.* **2021**, *261*, 117848. [CrossRef]

3. Shen, X.J.; Sun, R.C. Recent advances in lignocellulose prior-fractionation for biomaterials, biochemicals, and bioenergy. *Carbohydr. Polym.* **2021**, *261*, 117884. [CrossRef]

4. Liu, X.; Huang, K.X.; Lin, X.X.; Li, H.X.; Tao, T.; Wu, Q.H.; Zheng, Q.H.; Huang, L.L.; Ni, Y.H.; Chen, L.H.; et al. Transparent and conductive cellulose film by controllably growing aluminum doped zinc oxide on regenerated cellulose film. *Cellulose* **2020**, *27*, 4847–4855. [CrossRef]

5. Yousefi, H.; Faezipour, M.; Hedjazi, S.; Mousavi, M.M.; Azusa, Y.; Heidari, A.H. Comparative study of paper and nanopaper properties prepared from bacterial cellulose nanofibers and fibers/ground cellulose nanofibers of canola straw. *Ind. Crop. Prod.* **2013**, *43*, 732–737. [CrossRef]

6. Klemm, D.; Heublein, B.; Fink, H.P.; Bohn, A. Cellulose: Fascinating biopolymer and sustainable raw material. *Angew. Chem. Int. Ed. Engl.* **2005**, *44*, 3358–3393. [CrossRef] [PubMed]

7. Zhang, H.; Wu, J.; Zhang, J.; He, J. 1-allyl-3-methylimidazolium chloride room temperature ionic liquid a new and powerful nonderivatizing solvent for cellulose. *Macromolecules* **2005**, *38*, 8272–8277. [CrossRef]

8. Protz, R.; Lehmann, A.; Ganster, J.; Fink, H.P. Solubility and spinnability of cellulose-lignin blends in aqueous nmmo. *Carbohydr. Polym.* **2021**, *251*, 117027. [CrossRef]

9. Jadhav, S.; Lidhure, A.; Thakre, S.; Ganvir, V. Modified lyocell process to improve dissolution of cellulosic pulp and pulp blends in nmmo solvent. *Cellulose* **2021**, *28*, 973–990. [CrossRef]

10. Yasin, S.; Hussain, M.; Zheng, Q.; Song, Y.H. Effects of ionic liquid on cellulosic nanofiller filled natural rubber bionanocomposites. *J. Colloid Interface Sci.* **2021**, *591*, 409–417. [CrossRef]

11. Chen, F.; Bouvard, J.L.; Sawada, D.; Pradille, C.; Hummel, M.; Sixta, H.; Budtova, T. Exploring digital image correlation technique for the analysis of the tensile properties of all-cellulose composites. *Cellulose* **2021**, *28*, 4165–4178. [CrossRef]

12. Menezes, D.B.; Diz, F.M.; Ferreira, L.F.R.; Corrales, Y.; Baudrit, J.R.V.; Costa, L.P.; Hernandez-Macedo, M.L. Starch-based biocomposite membrane reinforced by orange bagasse cellulose nanofibers extracted from ionic liquid treatment. *Cellulose* **2021**, *28*, 4137–4149. [CrossRef]

13. Mamontov, E.; Osti, N.C.; Ryder, M.R. Order-disorder in room-temperature ionic liquids probed via methyl quantum tunneling. *Struct. Dyn.* **2021**, *8*, 024303. [CrossRef] [PubMed]

14. Phadagi, R.; Singh, S.; Hashemi, H.; Kaya, S.; Venkatesu, P.; Ramjugernath, D.; Ebenso, E.E.; Bahadur, I. Understanding the role of dimethylformamide as co-solvents in the dissolution of cellulose in ionic liquids: Experimental and theoretical approach. *J. Mol. Liq.* **2021**, *328*, 115392. [CrossRef]

15. Wu, J.; Zhang, J.; Zhang, H.; He, J.; Ren, Q.; Guo, M. Homogeneous acetylation of cellulose in a new ionic liquid. *Biomacromolecules* **2004**, *5*, 266–268. [CrossRef]

16. Song, H.Z.; Zhang, J.; Niu, Y.H.; Wang, Z.G. Phase transition and rheological behaviors of concentrated cellulose/ionic liquid solutions. *J. Phys. Chem. B* **2010**, *114*, 6006–6013. [CrossRef]

17. Bendaoud, A.; Kehrbusch, R.; Baranov, A.; Duchemin, B.; Maigret, J.E.; Falourd, X.; Staiger, M.P.; Cathala, B.; Lourdin, D.; Leroy, E. Nanostructured cellulose-xyloglucan blends via ionic liquid/water processing. *Carbohydr. Polym.* **2017**, *168*, 163–172. [CrossRef] [PubMed]

18. Gabrys, T.; Fryczkowska, B.; Binias, D.; Slusarczyk, C.; Fabia, J. Preparation and properties of composite cellulose fibres with the addition of graphene oxide. *Carbohydr. Polym.* **2021**, *254*, 117436. [CrossRef]

19. Cai, T.; Zhang, H.H.; Guo, Q.H.; Shao, H.L.; Hu, X.C. Structure and properties of cellulose fibers from ionic liquids. *J. Appl. Polym. Sci.* **2010**, *115*, 1047–1053. [CrossRef]

20. Stark, N.M. Opportunities for cellulose nanomaterials in packaging films: A review and future trends. *J. Renew. Mater.* **2016**, *4*, 313–326. [CrossRef]

21. Neves, R.M.; Ornaghi, H.L.; Zattera, A.J.; Amico, S.C. Recent studies on modified cellulose/nanocellulose epoxy composites: A systematic review. *Carbohydr. Polym.* **2021**, *255*, 117366. [CrossRef]

22. Liu, J.X.; Chen, P.; Qin, D.J.; Jia, S.; Jia, C.; Li, L.; Bian, H.L.; Wei, J.; Shao, Z.Q. Nanocomposites membranes from cellulose nanofibers, sio2 and carboxymethyl cellulose with improved properties. *Carbohydr. Polym.* **2020**, *233*, 115818. [CrossRef]

23. Song, H.Z.; Luo, Z.Q.; Wang, C.Z.; Hao, X.F.; Gao, J.G. Preparation and characterization of bionanocomposite fiber based on cellulose and nano-sio_2 using ionic liquid. *Carbohydr. Polym.* **2013**, *98*, 161–167. [CrossRef] [PubMed]
24. Raabe, J.; Fonseca, A.D.; Bufalino, L.; Ribeiro, C.; Martins, M.A.; Marconcini, J.M.; Tonoli, G.H.D. Evaluation of reaction factors for deposition of silica (sio_2) nanoparticles on cellulose fibers. *Carbohydr. Polym.* **2014**, *114*, 424–431. [CrossRef] [PubMed]
25. Lee, J.A.; Yoon, M.J.; Lee, E.S.; Lim, D.Y.; Kim, K.Y. Preparation and characterization of cellulose nanofibers (cnfs) from microcrystalline cellulose (mcc) and cnf/polyamide 6 composites. *Macromol. Res.* **2014**, *22*, 738–745. [CrossRef]
26. He, M.; Duan, B.; Xu, D.F.; Zhang, L.N. Moisture and solvent responsive cellulose/sio2 nanocomposite materials. *Cellulose* **2015**, *22*, 553–563. [CrossRef]
27. Tang, C.; Zhang, S.; Wang, X.B.; Hao, J. Enhanced mechanical properties and thermal stability of cellulose insulation paper achieved by doping with melamine-grafted nano-sio_2. *Cellulose* **2018**, *25*, 3619–3633. [CrossRef]
28. Kim, U.J.; Kimura, S.; Wada, M. Facile preparation of cellulose-sio_2 composite aerogels with high sio_2 contents using a libr aqueous solution. *Carbohydr. Polym.* **2019**, *222*, 114975. [CrossRef]
29. Segal, L.J.J.C.; Martin, A.E.J.; Conrad, C.M. An empirical method for estimating the degree of crystallinity of native cellulose using the x-ray diffractometer. *Text. Res. J.* **1959**, *29*, 786–794. [CrossRef]
30. Olsson, C.; Westman, G. Wet spinning of cellulose from ionic liquid solutions-viscometry and mechanical performance. *J. Appl. Polym. Sci.* **2013**, *127*, 4542–4548. [CrossRef]
31. Liu, L.; Ju, M.T.; Li, W.Z.; Hou, Q.D. Dissolution of cellulose from afex-pretreated zoysia japonica in amimcl with ultrasonic vibration. *Carbohydr. Polym.* **2013**, *98*, 412–420. [CrossRef]
32. Xia, G.M.; Wan, J.Q.; Zhang, J.M.; Zhang, X.Y.; Xu, L.L.; Wu, J.; He, J.S.; Zhang, J. Cellulose-based films prepared directly from waste newspapers via an ionic liquid. *Carbohydr. Polym.* **2016**, *151*, 223–229. [CrossRef]
33. Reddy, J.P.; Rajulu, A.V.; Rhim, J.W.; Seo, J. Mechanical, thermal, and water vapor barrier properties of regenerated cellulose/nano-sio2 composite films. *Cellulose* **2018**, *25*, 7153–7165. [CrossRef]
34. Arthanareeswaran, G.; Thanikaivelan, P.; Srinivasn, K.; Mohan, D.; Rajendran, M. Synthesis, characterization and thermal studies on cellulose acetate membranes with additive. *Eur. Polym. J.* **2004**, *40*, 2153–2159. [CrossRef]
35. Dlomo, K.; Mohomane, S.M.; Motaung, T.E. Influence of silica nanoparticles on the properties of cellulose composite membranes: A current review. *Cell Chem. Technol.* **2020**, *54*, 765–775. [CrossRef]
36. Nelson, M.L.; O'Connor, R.T. Relation of certain infrared bands to cellulose crystallinity and crystal lattice type. Part ii. A new infrared ratio for estimation of crystallinity in cellulose i and ii. *J. Appl. Polym. Sci.* **1964**, *8*, 1325–1341. [CrossRef]
37. Langan, P.; Nishiyama, Y.; Chanzy, H. A revised structure and hydrogen-bonding system in cellulose ii from a neutron fiber diffraction analysis. *J. Am. Chem. Soc.* **1999**, *121*, 9940–9946. [CrossRef]
38. Ashori, A.; Sheykhnazari, S.; Tabarsa, T.; Shakeri, A.; Golalipour, M. Bacterial cellulose/silica nanocomposites: Preparation and characterization. *Carbohydr. Polym.* **2012**, *90*, 413–418. [CrossRef] [PubMed]
39. Zhu, Y.C.; Wu, C.J.; Yu, D.M.; Ding, Q.J.; Li, R.G. Tunable micro-structure of dissolving pulp-based cellulose nanofibrils with facile prehydrolysis process. *Cellulose* **2021**, *28*, 3759–3773. [CrossRef]
40. Vallejo, M.; Cordeiro, R.; Dias, P.A.N.; Moura, C.; Henriques, M.; Seabra, I.J.; Malca, C.M.; Morouco, P. Recovery and evaluation of cellulose from agroindustrial residues of corn, grape, pomegranate, strawberry-tree fruit and fava. *Bioresour. Bioprocess.* **2021**, *8*, 25. [CrossRef]
41. Liu, Y.; Jing, S.; Carvalho, D.; Fu, J.; Martins, M.; Cavaco-Paulo, A. Cellulose dissolved in ionic liquids for modification of the shape of keratin fibers. *ACS Sustain. Chem. Eng.* **2021**, *9*, 4102–4110. [CrossRef]
42. Bai, L.M.; Liu, Y.T.; Ding, A.; Ren, N.Q.; Li, G.B.; Liang, H. Fabrication and characterization of thin-film composite (tfc) nanofiltration membranes incorporated with cellulose nanocrystals (cncs) for enhanced desalination performance and dye removal. *Chem. Eng. J.* **2019**, *358*, 1519–1528. [CrossRef]
43. Xie, K.; Yu, Y.; Shi, Y. Synthesis and characterization of cellulose/silica hybrid materials with chemical crosslinking. *Carbohydr. Polym.* **2009**, *78*, 799–805. [CrossRef]
44. Han, D.L.; Yan, L.F.; Chen, W.F.; Li, W.; Bangal, P.R. Cellulose/graphite oxide composite films with improved mechanical properties over a wide range of temperature. *Carbohydr. Polym.* **2011**, *83*, 966–972. [CrossRef]
45. Yang, Y.P.; Zhang, Y.; Dawelbeit, A.; Deng, Y.; Lang, Y.X.; Yu, M.H. Structure and properties of regenerated cellulose fibers from aqueous naoh/thiourea/urea solution. *Cellulose* **2017**, *24*, 4123–4137. [CrossRef]
46. Yang, Y.S.; Shan, L.; Shen, H.J.; Qiu, J. Green and facile fabrication of thermal superamphiphobic nanofibrillated-cellulose/chitosan/ots composites through mechano-chemical method. *Fibers Polym.* **2021**, *22*, 1407–1415. [CrossRef]
47. Bunmechimma, L.; Leejarkpai, T.; Riyajan, S.A. Fabrication and physical properties of a novel macroporous poly(vinyl alcohol)/cellulose fibre product. *Carbohydr. Polym.* **2020**, *240*, 116215. [CrossRef]
48. Okon, K.E.; Lin, F.C.; Chen, Y.D.; Huang, B. Effect of silicone oil heat treatment on the chemical composition, cellulose crystalline structure and contact angle of chinese parasol wood. *Carbohydr. Polym.* **2017**, *164*, 179–185. [CrossRef] [PubMed]
49. Collier, J.R.; Watson, J.L.; Collier, B.J.; Petrovan, S. Rheology of 1-butyl-3-methylimidazolium chloride cellulose solutions. Ii. Solution character and preparation. *J. Appl. Polym. Sci.* **2009**, *111*, 1019–1027. [CrossRef]
50. Gericke, M.; Schlufter, K.; Liebert, T.; Heinze, T.; Budtova, T. Rheological properties of cellulose/ionic liquid solutions: From dilute to concentrated states. *Biomacromolecules* **2009**, *10*, 1188–1194. [CrossRef]

51. Yao, Y.B.; Yan, Z.Y.; Li, Z.; Zhang, Y.M.; Wang, H.P. Viscoelastic behavior and sol-gel transition of cellulose/silk fibroin/1-butyl-3-methylimidazolium chloride extended from dilute to concentrated solutions. *Polym. Eng. Sci.* **2018**, *58*, 1931–1936. [CrossRef]
52. Yao, Y.B.; Xia, X.L.; Mukuze, K.S.; Zhang, Y.M.; Wang, H.P. Study on the temperature-induced sol-gel transition of cellulose/silk fibroin blends in 1-butyl-3-methylimidazolium chloride via rheological behavior. *Cellulose* **2014**, *21*, 3737–3743. [CrossRef]
53. Shakeel, A.; Mahmood, H.; Farooq, U.; Ullah, Z.; Yasin, S.; Iqbal, T.; Chassagne, C.; Moniruzzaman, M. Rheology of pure ionic liquids and their complex fluids: A review. *ACS Sustain. Chem. Eng.* **2019**, *7*, 13586–13626. [CrossRef]
54. Yao, Y.B.; Mukuze, K.S.; Zhang, Y.M.; Wang, H.P. Rheological behavior of cellulose/silk fibroin blend solutions with ionic liquid as solvent. *Cellulose* **2014**, *21*, 675–684. [CrossRef]
55. Xia, X.L.; Wang, J.N.; Wang, H.P.; Zhang, Y.M. Numerical investigation of spinneret geometric effect on spinning dynamics of dry-jet wet-spinning of cellulose/ bmim cl solution. *J. Appl. Polym. Sci.* **2016**, *133*, 49362. [CrossRef]
56. Rajeev, A.; Deshpande, A.P.; Basavaraj, M.G. Rheology and microstructure of concentrated microcrystalline cellulose (mcc)/1-allyl-3-methylimidazolium chloride (amimcl)/water mixtures. *Soft Matter* **2018**, *14*, 7615–7624. [CrossRef]

nanomaterials

Article

5-Fluorouracil Encapsulated Chitosan-Cellulose Fiber Bionanocomposites: Synthesis, Characterization and In Vitro Analysis towards Colorectal Cancer Cells

Mostafa Yusefi [1], Hui-Yin Chan [2], Sin-Yeang Teow [2], Pooneh Kia [3], Michiele Lee-Kiun Soon [2], Nor Azwadi Bin Che Sidik [1,*] and Kamyar Shameli [1,*]

1   Malaysia-Japan International Institute of Technology, Universiti Teknologi Malaysia, Jalan Sultan Yahya Petra, Kuala Lumpur 54100, Malaysia; myusefi175@gmail.com
2   Department of Medical Sciences, School of Medical and Life Sciences, Sunway University, Jalan Universiti, Bandar Sunway, Selangor Darul Ehsan 47500, Malaysia; 19117928@imail.sunway.edu.my (H.-Y.C.); ronaldt@sunway.edu.my (S.-Y.T.); 19117936@imail.sunway.edu.my (M.L.-K.S.)
3   Institute of Bio Science, University Putra Malaysia, Serdang 43400, Malaysia; kia.pooneh@gmail.com
*   Correspondence: azwadi@utm.my (N.A.B.C.S.); kamyarshameli@gmail.com (K.S.)

Citation: Yusefi, M.; Chan, H.-Y.; Teow, S.-Y.; Kia, P.; Lee-Kiun Soon, M.; Sidik, N.A.B.C.; Shameli, K. 5-Fluorouracil Encapsulated Chitosan-Cellulose Fiber Bionanocomposites: Synthesis, Characterization and In Vitro Analysis towards Colorectal Cancer Cells. *Nanomaterials* **2021**, *11*, 1691. https://doi.org/10.3390/nano11071691

Academic Editors: Carla Vilela, Carmen S. R. Freire and Eleonore Fröhlich

Received: 4 April 2021
Accepted: 19 May 2021
Published: 28 June 2021

Publisher's Note: MDPI stays neutral with regard to jurisdictional claims in published maps and institutional affiliations.

Abstract: Cellulose and chitosan with remarkable biocompatibility and sophisticated physiochemical characteristics can be a new dawn to the advanced drug nano-carriers in cancer treatment. This study aims to synthesize layer-by-layer bionanocomposites from chitosan and rice straw cellulose encapsulated 5-Fluorouracil (CS-CF/5FU BNCs) using the ionic gelation method and the sodium tripolyphosphate (TPP) cross-linker. Data from X-ray and Fourier-transform infrared spectroscopy showed successful preparation of CS-CF/5FU BNCs. Based on images of scanning electron microscopy, 48.73 ± 1.52 nm was estimated for an average size of the bionanocomposites as spherical chitosan nanoparticles mostly coated rod-shaped cellulose reinforcement. 5-Fluorouracil indicated an increase in thermal stability after its encapsulation in the bionanocomposites. The drug encapsulation efficiency was found to be $86 \pm 2.75\%$. CS-CF/5FU BNCs triggered higher drug release in a media simulating the colorectal fluid with pH 7.4 ($76.82 \pm 1.29\%$) than the gastric fluid with pH 1.2 ($42.37 \pm 0.43\%$). In in vitro cytotoxicity assays, cellulose fibers, chitosan nanoparticles and the bionanocomposites indicated biocompatibility towards CCD112 normal cells. Most promisingly, CS-CF/5FU BNCs at 250 µg/mL concentration eliminated $56.42 \pm 0.41\%$ of HCT116 cancer cells and only $8.16 \pm 2.11\%$ of CCD112 normal cells. Therefore, this study demonstrates that CS-CF/5FU BNCs can be considered as an eco-friendly and innovative nanodrug candidate for potential colorectal cancer treatment.

Keywords: cellulose; chitosan nanoparticles; bionanocomposites; 5-fluorouracil; in vitro drug release; cytotoxicity assay; colorectal cancer

1. Introduction

Currently, conversion of natural-based residues into useful and novel materials may tackle financial and environmental issues [1]. In this regard, rice straw is an abundant lignocellulosic residue, which contains a high ratio of cellulose to use in a myriad of research fields [2,3]. As a nature gifted material, cellulose fiber is an important component of the wood cell wall and the most abundant organic polymer on earth [4]. Plant cellulose possesses advantageous and unique mechanical, optical and rheological characteristics, along with a sensitivity to the particular molecular structure of the antigen and pH-sensitivity for synthesis of novel polymeric nanodrug formulations [4]. The cellulose, however, does possess some drawbacks, including poor crease resistance and low solubility in solvent fluids. Its physiochemical characteristics suitably can be modified by esterification, etherification, de-polymerization, radical grafting and alkali treatments to impart exogenous groups over the cellulose structure without damaging its advantageous intrinsic characteristics [5].

Chitosan is the second most popular biopolymer after cellulose, with production of over 100 million tons per year [6]. It may be derived from chitin and is a cationic linear and natural amino-polysaccharide containing -(1-4)-linked d-glucosamine and *N*-acetyld-glucosamine in deacetylated and acetylated form, respectively [7]. Among diverse methods to synthesize layer-by-layer chitosan-based composites and nanoparticles (NPs), ionic gelation approach is an organic solvent-free solution, straightforward and a facile method with minimal toxicity [8]. In this method, the phosphate groups of sodium tripolyphosphate (TPP) may act as a physical crosslinking agent, which has advantages over emulsifying and chemical crosslinking agents, such as less toxicity to the organs and no destruction to the structure of the loaded-drugs in chitosan nanoparticles [9]. Polymer blends are increasingly important to fabricate innovative composites for various applications [10]. The blend of degradable polymers can merge the desirable properties of the polymers [11]. In addition, the crosslinking procedure might considerably improve physiochemical properties of the polymer composites [12]. In medically-related applications, the most popular antimicrobial coating agent on cellulose is currently chitosan to synthesize composites with suitable biocompatibility and water-rich structures to encapsulate both hydrophilic and hydrophobic drugs [13]. Furthermore, the bionanocomposites or nanohybrids of chitosan and cellulose possess intermolecular interactions, owing to H-bonds and Van der Waals forces [14]. Most importantly, chitosan-cellulose bionanocomposites or nanohybrids may possess a tremendous swelling capacity and water absorption ability to release the drug at the targeted cells [15]. Therefore, the biocompatibility and physiochemical properties of both cellulose and chitosan can be modified by using cellulose as a reinforcement and chitosan as a coating agent to construct a double polysaccharide composite biomaterial.

Colorectal cancer (CRC) is the third most diagnosed cancer with death of six lakhs per year [16]. 5-Fluorouracil (5FU) as an antimetabolite and fluorinated pyrimidine has been used for CRC therapy since three decades ago [17]. Albeit, the unwanted side-effects of using 5FU-based chemotherapy might cause diarrhea, stomatitis and gastrointestinal mucosal injury. The above side-effects can be abated by encapsulation of a minimal drug dosage into a biocompatible drug carrier. Of this, 5FU has been encapsulated in many drug carrier platform, such as, chitosan [18] and cellulose [19,20]. In this matter, some polysaccharides, including cellulose, chitosan, starch, xanthan gum, hyaluronic acid, and carrageenans can display pH/thermos-responsive properties to control the drug release dosage with effective anticancer actions [21].

Nano-sized medicines may favorably enhance circulation time, targeted activities, and safe drug delivery to decrease the issues of side effects from the chemotherapy treatments [22–24]. For this aim, biopolymer-based NPs may ideally avoid the immune system to increase circulation time of the NPs in blood to enhance the anticancer actions [25]. The drug delivery systems may affect the microenvironment of tumors, which is leaky and has a higher sensitivity to macromolecules compared to the normal cells [26]. Thus, drug encapsulation in natural polymers can be considered as an emerging research area for different pharmaceutical applications [24,25,27]. Furthermore, being impressively biodegradable and possessing an ability for the enzymatic degradation by colonic microbial agents, biopolymers (such as lignin-based NPs, glycogen and gelatin and cellulose fibers) may trigger probes for novel polymeric–anticancer nanodrug conjugates with advanced antiproliferative actions [24,25,27,28].

For example, 5FU-loaded composites of nanocellulose (obtained from commercial α-cellulose), chitosan and sodium alginate eliminated HT29 CRC cells, however, it unfavorably killed the normal HEK293 cells at concentrations between 50 to 100 μg/mL [14]. In a separate experiment, curcumin loaded nanocellulose caused three times higher effect compared to curcumin alone against HT29 CRC cells, albeit, the biocompatibility of the sample was not studied [29]. In recent investigations, the release of 5FU was controlled through nanocellulose/gelatin composites [30] and also 3D printed composites of cellulose nanofibrils and $CaCO_3$ [31]. Both studies did not examine the anticancer action of the samples.

In this present study, cellulose fibers were extracted from rice straw waste. In addition, chitosan NPs and bionanocomposites of chitosan-cellulose were fabricated by the ionic gelation method using TPP as a cross-linker. Above all, 5FU was encapsulated in the bionanocomposites of chitosan as a coating agent and cellulose as a reinforcement for CRC analysis. To the best of our knowledge, this is the first time that rice straw cellulose was coated with chitosan for anticancer drug 5FU carrier in potential CRC treatments. The physiochemical properties of the synthesized samples were evaluated by X-ray powder diffraction (XRD), scanning electron microscopy (SEM), energy dispersion X-ray spectroscopy (EDX), thermogravimetric analysis (TGA), dynamic light scattering (DLS), Fourier-transform infrared spectroscopy (FTIR), and the swelling analysis. The drug absorbance and release of 5FU-loaded in the bionanocomposites were evaluated by using ultraviolet–visible (UV) spectroscopy. In vitro cytotoxicity assays evaluated the biocompatibility and anticancer activity of the fabricated samples against CCD112 colon normal and HCT116 CRC cell lines.

2. Materials and Methods

2.1. Materials

Rice straw waste was obtained from Malaysian Agricultural Research and Development Institute (MARDI), Selangor, Malaysia. Potassium hydroxide (KOH, 85%), sodium chlorite ($NaClO_2$, 80%), acetic acid glacial (CH_3COOH) (98%), chitosan (low molecular weight, 190,000–310,000 degree of acetylation), Tween-80 and TPP were all purchased from Sigma Aldrich (St. Louis, MO, USA). 5FU, 99%, 5-Fluoro-2,4(1H,3H)-pyrimidinedione (ACD CODE MFC D00006018) with a molecular weight of 130.08 g/mol was purchased from ACROS ORGANICS part of Thermo Fisher Scientific, Branchburg, NJ, USA. The chemicals were used without further purification. All glassware used was washed with distilled water and dried before used.

2.2. Synthesis

2.2.1. Extraction of Cellulose Fibers from Rice Straw Waste

CF was extracted from rice straw waste through a series of chemical treatments. The rice straw powder (30 g) was dewaxed by using a soxhlet instrument as it contained a 450 mL solution of toluene/ethanol 2:1 (v/v) for 12 h at 70 °C. After that, the sample was washed with distilled water three times and mixed with a sodium chloride solution (1.4%), in which dropwise adding acetic acid adjusted the pH to around 4 at 70 °C under 100 rpm magnetic stirring for 5 h. The sample was washed three times and then treated with KOH solution (5%) for 12 h, followed by pouring 10-fold ice cubes into the sample solution to obtain cellulose. The cellulose solution was washed and centrifuged (Tabletop Centrifuge Kubota, Model: 2420) three times at 4000 rpm for 8 min. Finally, the suspension was freeze-dried (FreeZone 1.0 L Benchtop Freeze Dry System, Kansas City, MO, USA) at −45 °C for 48 h and termed as CF, which stored at −4 °C for further analysis.

2.2.2. Synthesis of Chitosan-Cellulose Fiber Bionanocomposites to Encapsulate 5-Fluorouracil

Ionic gelation method was performed to synthesize layer-by-layer of three chitosan-based samples: (i) chitosan-cellulose fibers bionanocomposites encapsulated 5FU (CS-CF/5FU BNCs), (ii) chitosan-cellulose fibers bionanocomposites (CS-CF BNCs), and (iii) chitosan nanoparticles (CS NPs). First, three 250 mL beakers respectively contained 80 mL mixture solution of 1.0% acetic acid and 0.250 g of chitosan powder (low molecular weight). Then, 2% (v/v) of Tween-80 as a stabilizer was respectively added to each solution and mixed gently for 45 min to obtain the three chitosan solutions. To prepare CS-CF/5FU BNCs, 0.125 g CF and 0.01 g 5FU were mixed with one of the prepared chitosan solutions and then homogenized vigorously (DAIHAN-brand Homogenizer, Wonju-Si, Gang-Won -Do, Korea) at 9000 rpm for 7 min. After that, 0.50 g of TPP cross-linker was dissolved in 15 mL deionized water and added dropwise to the CS-CF/5FU solution under the continuously vigorous stirring of the homogenizer for another 45 min. The mixture solution was washed with distilled water and centrifuged three times at 2500 rpm for 7 min at 25 °C. Finally, the sample was freeze-dried for 16 h and stored at −4 °C for further analysis.

To prepare CS-CF BNCs, the above procedure was performed, however, without adding 5FU to the CS-CF solution.

To prepare CS NPs, the above method was used, however, without adding CF and 5FU to the CS solution.

2.3. Characterization

2.3.1. Physicochemical Analysis

XRD (Philips, X'pert, Cu Ka, Amsterdam, North Holland, The Netherlands) at an ambient condition was used to evaluate the structure of the samples. The sample was compressed between two smooth glass films and the XRD analysis was carried out in dispersion 2 angles of 5°–80° at a step size of 0.02° with 2 s/step as scanning rate using a voltage of 45 kV, a Ni-filtered Cu K radiation (=1.5406 A)° and a filament current of 40 mA. The SEM images were taken via using an Electro-Scan SEM instrument (model JSM 7600 F SEM, Tokyo, Kantō, Japan) attached to EDX to study the elemental composition of the sample. A low-acceleration voltage (10 kV) was used to prevent the degradation of the sample. Thermal analysis was carried out though using TGA (STA F3 Jupiter, Selb, Bavaria, Germany) Q50 V20 at a heating rate of 10 °C/min under a nitrogen atmosphere (10 mL/min) from 10 °C to 800 °C. For DLS analysis, an Anton Paar instrument (Litesizer 500, Graz, Styria, Austria) was used to measure the hydrodynamic size of the synthesized samples in buffer solutions (100 µg/mL) at various pH values of 1.2, 7.4 and 12 at 37 °C. Hydrochloric acid (HCl) and sodium hydroxide (NaOH) were used to adjust the pH 1.2 and 12, respectively. FTIR spectroscopy (ThermoNicolet, Waltham, MA, USA) determined the chemical and super-molecular structural analysis of the samples under an ambient condition. First, crushing and mixing of the sample with KBr at a ratio of 1:100 w/w to prepare a transparent pellet and the spectra of the plate was evaluated under a transmittance mode in a range between 4000 cm^{-1} to 400 cm^{-1} with a 4 cm^{-1} resolution and an accumulation of 128 scans. The swelling ratio of the sample was measured. For this aim, the sample was immersed in the solutions with various pH values of 1.2, 7.4, and 12 at 37 °C under 75 rpm magnetic stirring [32]. At each interval, the sample was collected from the solution and blotted on a filter paper to eliminate excess water and it was immediately weighed to determine the weight of the wet sample. The swelling ratio of the sample was measured as W_t/W_0, in which W_t and W_0 are the obtained wet weights of the sample at an arbitrary and initial time, respectively. For the DLS analysis and swelling experiment, individual tests were repeated three times, which the data were indicated as mean ± standard deviation for all triplicates within an independent test.

2.3.2. Encapsulation Efficiency Study of 5-Fluorouracil

UV–vis spectrophotometry (UV-1600, Shimadzu, Kyoto, Kansai, Japan) provided a calibration curve at λ_{max} = 266 nm from known concentrations of 5FU (5–25 ppm). This followed by calculating the amount of the drug-loaded into CS-CF/5FU BNCs by using Equation (1) to obtain drug encapsulation efficiency% [33].

$$\text{Encapsulation efficiency } (\%) = \frac{(\text{Initial drug amount in formulation (mg)} - \text{Unentrapped drug (mg)})}{\text{Initial drug amount in formulation (mg)}} \times 100 \quad (1)$$

2.3.3. A Comparative Study of In Vitro Release of 5-Fluorouracil Drug from CS-CF/5FU BNCs

For the drug release performance of CS-CF/5FU BNCs, a 5 mL dialysis bag (molecular weight cut-off between 12,000 and 14,000 Da) was used according to reported studies with slight modification [20,34]. Before the experiment, the bag was soaked for 12 h in the release media of the simulated colorectal fluid (phosphate-buffered saline (PBS) at pH 7.4).

Then, the dialysis bag with the two ends tied as contained the solution mixture of 5 mg of CS-CF/5FU BNCs and 2 mL of the release media. The bag was completely immersed into a 40 mL of the release media maintained under constant stirring of 100 rpm at 37 °C in the stoppered bottle. A 1 mL aliquot was withdrawn from the system at the selected time and characterized by UV–vis spectrophotometry at λ_{max} = 266 nm. The same study was performed in HCl buffer solution at pH 1.2 for the simulated release of the 5-FU drug in the gastric fluid without enzymes. The following Equation (2) calculated and compared the drug release results from the fluids with the two different pH values:

$$\text{Drug release (\%)} = \frac{\text{Released drug at time } 't'}{\text{Total drug in the sample (CS} - \text{CF/5FU BNCs)}} \times 100 \qquad (2)$$

2.3.4. Cell Lines and Reagents

Human HCT116 colorectal cancer (CRC) (ATCC CCL-247) and CCD112 normal colon (ATCC CRL-1541) cell lines were purchased from ATCC (VA, USA) and cultured according to ATCC's recommendation [35,36]. The cell lines were maintained in Dulbecco's Modified Eagle's medium (DMEM) supplemented with 10% fetal bovine serum (FBS) (Gibco) and 1% penicillin/streptomycin (Gibco) under standard cell culture conditions.

2.3.5. In Vitro Cytotoxicity Assay

In vitro cytotoxicity assays were performed using CellTiter 96 Aqueous One Solution kit (MTS reagent) (#G3582, Promega), according to the manufacturer's instruction with slight modification as previously described [37]. Briefly, 5000 HCT116 and CCD112 cells per well (100 µL/well) were separately seeded onto a 96 well plate and incubated overnight at 37 °C in a 95% humidified incubator with 5% CO_2. The next day, 2-fold serially diluted samples at concentrations of 0, 7.81, 15.62, 31.25, 62.53, 125, 250 and 500 (100 µL/well) were added into the wells and the plate was incubated for 72 h at 37 °C in the 5% CO_2 incubator. Then, 20 µL of the MTS reagent per well was added into the plate and incubated for an additional 3 h at 37 °C in the 5% CO_2 incubator. The optical density (OD) was then measured at 490 nm using a multimode microplate reader (Tecan). The dose–response graph was plotted by calculating the percent cell viability using Equation (3) [38]:

$$\text{\% Cell viability} = \frac{\text{OD of sample well (mean)}}{\text{OD of control well (mean)}} \times 100 \qquad (3)$$

The inhibitory concentration causing 50% growth inhibition (IC_{50}) was determined through an online calculator (https://www.aatbio.com/tools/ic50-calculator (accessed on 5 May 2021)) as previously described [35].

2.3.6. Statistical Analysis

Independent experiments were performed three times and data were expressed as mean ± standard deviation for all triplicates within an individual experiment. Data were analyzed with a Student's *t* test using SPSS version 26.0. $p < 0.05$ was considered significant.

3. Results and Discussion

Scheme 1 depicts the synthesis of a novel nanocomposites of rice straw CF and CS encapsulated 5FU for the in vitro CRC analysis. CF was successfully extracted from rice straw waste by a series of chemical modifications, such as, bleaching, delignification, and alkali treatments. Sodium chloride solution and potassium hydroxide degraded lignin and hemicelluloses of the rice straw waste, respectively [39]. Then, the delignificated rice straw waste was treated with 5% potassium hydroxide to isolate CF containing nano-scale diameter and almost a uniform structure. This method was used by different studies, which found similar results from the extracted cellulose of the newspaper waste [40] and rice straw [41]. In a separate investigation [42], an increase in the ratio of potassium hydroxide from 5 to 15 weight% decreased the size of the abaca fibers; whereas, the 10 and 15 weight%

of the potassium hydroxide undesirably decreased the strength of fibers comprising a twisted structure. Thus, we used proper treatments of 5% potassium hydroxide solution to extract cellulose from the rice straw waste.

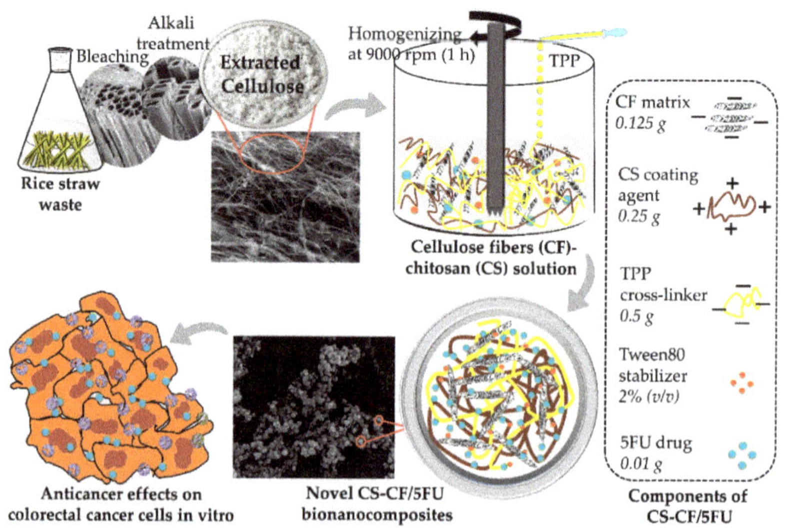

Scheme 1. A schematic process of synthesis of CS-CF/5FU BNCs for in vitro cytotoxicity assays towards CRC cell line.

Incorporation of dual polysaccharides of CF and CS in a layer-by-layer complex encapsulated the anticancer drug 5FU by using the ionic gelation method, where TPP, Tween-80, and acetic acid served as a cross-linking agent, stabilizer and hydrogen ion formation, respectively. The ionic gelation method is facile and responsible for the increased degree of cross-linking between the components of the CS-based composites [43]. CF as a reinforcement can be compatible with CS as a coating agent to synthesize dual biocompatible composites [7]. The characteristics of CS based complexes might be enhanced by reinforcement of cellulose. Also, CS as a coating agent could possibly show inherently biocompatibility, whereas its physiochemical characteristics could be enhanced by CF as a matrix to modulate the drug release and improve anticancer effects [44].

The ratio of the ingredients in the polymer composites is an important concern. The ratio between the chitosan powder to TPP cross-linker was 1:2 (*v/v*). Similarly, different research indicated that this ratio was suitable in which a reaction yield (75.0%), drug loading content (16.7%), and encapsulation efficiency (66.7%) were optimized to obtain chitosan NPs with desirable physiochemical properties. The study also used 1.0% acetic acid and 2% (*v/v*) Tween-80 in the chitosan solution. For the CS-CF composite, we use CF and chitosan with the ratio of 1:2 to obtain the spherical nanocomposite with size below 50 nm, as indicated in the SEM images. In a different study, 2.0% (*w/v*) chitosan, 1.2% (*w/v*) carboxymethyl-cellulose and 1.0% (*w/v*) scleroglucan were used to synthesize a nanocomposite hydrogels [45]. Based on SEM images, this sample was not as a spherical composite. In addition, the chitosan did not show homogeneous coating structure on the carboxymethyl-cellulose. This could be possibly due to using higher ratio of the cellulose and scleroglucan than chitosan. Studies by Samy et al. (2020) have shown that the ratio between chitosan and cellulose was 1:1 and epichlorohydrin acted as a cross-linker [46]. This composite showed size above 100 nm, according to the SEM images. In our study, therefore, the successful fabrication of CS-CF composites with size below 50 nm could be owing to blending the proper ratio of CS coating agent and CF reinforcement under high-speed homogenizing (9000 rpm), whereas TPP desirably acted as a crosslinking agent.

Further, the CS-CF composites may possess van der Waals interactions with the anticancer drug 5FU. For the CS-CF/5FU sample preparation, only 0.01 g 5FU was added in the CS-CF solution. It is worth to mention that using the low amount of drug in the nano-carrier system may cause high drug encapsulation efficiency and prolonged drug release to obtain effective anticancer action. Studies by Nguyen et al. (2017) showed that using the same amount of 5FU (0.01 g) in chitosan solution and ionic gelation method for the improved drug encapsulation and drug release [47]. Also, studies by Hosokawa et al. (1991) also showed that encapsulated the same amount of 5FU (0.01 g) with poly (ethylene glycol) methyl ether-chitosan nano-gels [48]. Therefore, the ratio of the ingredients in CS-CF/5FU could be suitable to obtain a nanodrug complex with relevant physiochemical properties and anticancer effects.

Scheme 2 indicates the possible intermolecular chemical interactions in the bio-nanocomposites. The presence of hydrogen and Van der Waals forces in CS-CF/5FU BNCs could trigger intermolecular interactions in the composites [14,19]. During the sample preparation, cationic groups of chitosan (NH_3^+) could react with the -OH^- and phosphoric ions in the TPP solution to support the deprotonation of CS. As reported [49], the formation of CS cross-linked CF could be obtained from Schiff base reaction between primary amino and carbonyl groups of CS and CF, respectively. Furthermore, 5FU was encapsulated and conjugated into the composites potentially by hydrogen bonding and van der Waals interactions [7]. The freeze-drying process was performed for the synthesized samples. It was reported that the freeze-drying process might trigger a gentle vaporization of the bound water and also sublimation of the ice crystals upon the samples to increase the presence of –OH groups and produce white fluffy fibrous materials [50].

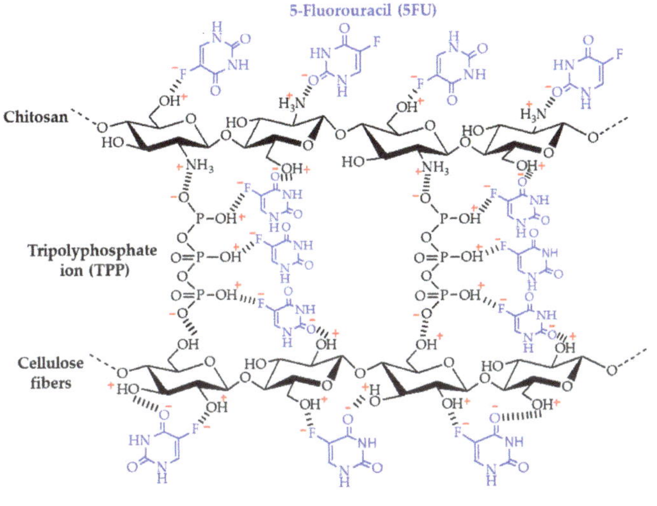

Scheme 2. The possible intermolecular chemical interactions between active functional groups in CS-CF/5FU BNCs.

3.1. Physicochemical Characterization of CS-CF/5FU BNCs Using X-ray Powder Diffraction

XRD was used to evaluate the crystallographic structure of the fabricated samples. Figure 1a–e show the XRD results of rice straw waste, CF, CS NPs, CS-CF BNCs, and CS-CF/5FU BNCs, respectively. The samples containing CF exhibited a similar XRD pattern, representing that the chemical treatments did not destroy the cellulose structure. The diffraction peaks of CF was approximately at 2θ = 15.51°, 22.23° and 35.32°, similar to the normal cellulose-I structure [39]. The main crystalline region was at 22.23° with a strong intensity, presenting the proper crystallinity of CF. Therefore, the delignification, bleaching as well as alkali treatments on the rice straw waste effectively degraded the amorphous regions and liberated the crystal regions. From Figure 1c, CS NPs presented no peaks of crystallinity since the crystalline structure of CS was eliminated after its crosslinking with TPP [51]. CS-CF BNCs and CS-CF/5FU BNCs performed a similar pattern attributed to both CF and CS peaks approximately at $2\theta = 22.23°$, $15.78°$ and $9.04°$, which is in a good agreement with the JCPDS card no. 04-0784 [52]. Noticeably, the peak at $15.78°$ is an overlapping peak between CF and CS. Compared to CF, the CS-CF/5FU BNCs sample showed lower crystallinity peaks, due to presence of the TPP cross-linker to abate the crystallinity [51]. Further results indicated that CS-CF/5FU BNCs did not have any sharp peak related to 5FU, since the drug was possibly entrapped within the carrier composites [53]. The 5FU alone could display a crystalline structure with a sharp diffraction peak at around $2\theta = 28.71°$ [53]. Moreover, the crystalline structure of the drug alone could possibly become a non-crystalline after its encapsulation within the carrier [54]. The XRD results indicated that CS-CF/5FU BNCs possessed a structure as polysaccharides to entrap 5FU.

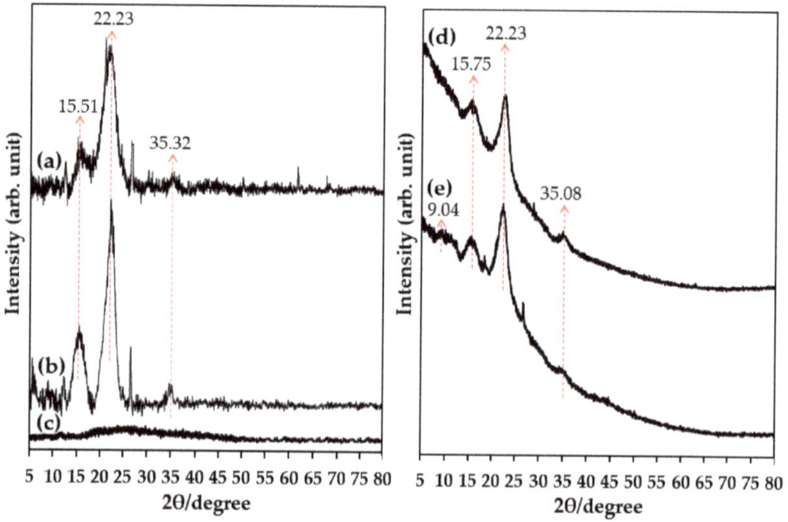

Figure 1. XRD spectra of (**a**) rice straw waste, (**b**) CF, (**c**) CS NPs, (**d**) CS-CF BNCs, (**e**) CS-CF/5FU BNCs.

3.2. Physicochemical Characterization of CS-CF/5FU BNCs Using Scanning Electron Microscopy and Energy Dispersive X-ray Analyses

SEM provides information about surface topography of the samples. Figure 2a–d show SEM images and EDX results for CF, CS NPs and CS-CF/5FU BNCs, respectively. The extracted CF was mostly in a rod-shaped structure comprised of individual and organized nano-fibrils with an average width of 50.81 ± 3.6 nm. The size of the rice straw waste gradually decreased after degradation of the hemicellulose and lignin through dewaxing, delignification and alkali treatments to liberate CF with nearly a uniform structure. CS NPs presented spherical shapes (24.17 ± 1.1 nm) and CS-CF/5FU BNCs demonstrated that CS NPs coated the CF network. An average diameter of CS-CF/5FU BNCs was found to be 48.73 ± 1.5 nm. The size of CS NPs is related to the ratio between the TPP and the chitosan powder. According to a separate report [55], the chitosan and TPP with a ratio around 1:2 could cause the formation of CS NPs with appropriate properties and nano-scale dimension, therefore, it was used in this current study. The use of the layer-by-layer synthesis potentially caused that CF was coated with CS and it was almost homogenously distributed in the multi-layered particles of the BNCs with low agglomeration. This was similarly found in different investigations [44,56]. It can be possibly noticed from two images of the composites that most of the rod-shaped CF was entangled in the spherical CS particles; thus, CS-CF/5FU BNCs mainly indicated the spherical shape. Whereas, a few CF assembled clusters of CS particles. The CF structure as a reinforcement could potentially trigger a positive influence to host 5FU and increase the drug conjugation and absorbance within the composite; moreover, the CS coating on CF possibly enhanced the drug entrapment and encapsulation in the composites.

It can be understood from the EDX results of CF that the chemical treatments on rice straw waste dissolved the silica in aqueous ions to be subsequently replaced by carbon. Compared to the EDX results of CF, the composites of CS-CF showed increasing and decreasing carbon and oxygen content, respectively (Figure 2a–c). A separate study stated that decreasing the amount of oxygen in CS-CF composites may increase the biodegradability of the CS-based composites [2]. As shown in EDX layered image (Figure 2d), CS-CF/5FU BNCs showed elements related to CS and CF as well as a negligible ratio of the fluorine (F) from 5FU. The above SEM and EDX results indicated that the nanocomposites successfully were synthesized to entrap anticancer drug 5FU.

3.3. Physicochemical Characterization of CS-CF/5FU BNCs Using Dynamic Light Scattering

DLS was used to estimate the size distribution of the particles in the solution. Figure 3a–c show the DLS results with values for hydrodynamic size of the synthesized samples in distilled water. The suspensions of CF, CS NPs, and CS-CF/5FU BNCs showed the hydrodynamic size of 162.58 ± 4.31, 133.18 ± 3.46, and 197.56 ± 4.12 nm, respectively. The obtained size was larger in DLS compared to that from SEM. This can be explained that DLS indicates the size of the particle and the surrounding diffuse layer of the particle, however, the SEM size is attributed to the particle itself. For the colloidal stability in the aqueous media, the presence of particle–particle interactions affected the hydrophobic attraction energy between the particles to attract each other.

The dimensional effect of the prepared nanofluid samples under various conditions was demonstrated. In order to deliver nanodrug complex into the target cells, the pH sensitivity of the fabricated samples is a vital concern. As depicted in Table 1, effects of pHs on hydrodynamic size of the nanofluid sample were evaluated using the samples at the solution with different pH values of 1.2 (strong acid), 7.4 (neutral) and 12 (strong base). All the synthesized samples indicated the size below 220 nm in the solution at the harsh conditions of strong acidic (pH 1.2) and basic (pH 12). This may possibly show that the harsh conditions did not damage the nano-dimensional of the examined samples. The results indicated that an increase in the pH of the CF solution from 1.2 to 12 slightly increased the hydrodynamic size of CF from 135.56 ± 2.87 nm to 203.17 ± 4.81 nm. Compared to CF, both CS-CF BNCs and CS-CF/5FU BNCs showed a higher enhancement of the size with increasing the pH value. This may address the pH-sensitive nature of chitosan as also stated in a different study [57]. At the solution with pH 1.2, CS NPs, CS-CF BNCs and CS-CF/5FU BNCs showed hydrodynamic size of 71.61 ± 5.54, 109.03 ± 4.12, and 112.51 ± 4.09 nm, respectively; whereas these samples indicated the size of 246.09 ± 5.10, 275.34 ± 4.59 and 274.23 ± 5.11 nm at the solution with pH 12. This increase in the hydrodynamic size of the samples could be attributed to the particle agglomeration. Particle aggregation might trigger reduced repulsive on the particles surface, due to the rise in the pH of the CF and CS solution [57]. It could be understood from the above results that the CS coating could enhance the pH sensitivity of the CF solid support to synthesize CS-CF/5FU BNCs as a pH-sensitive nanodrug formulation.

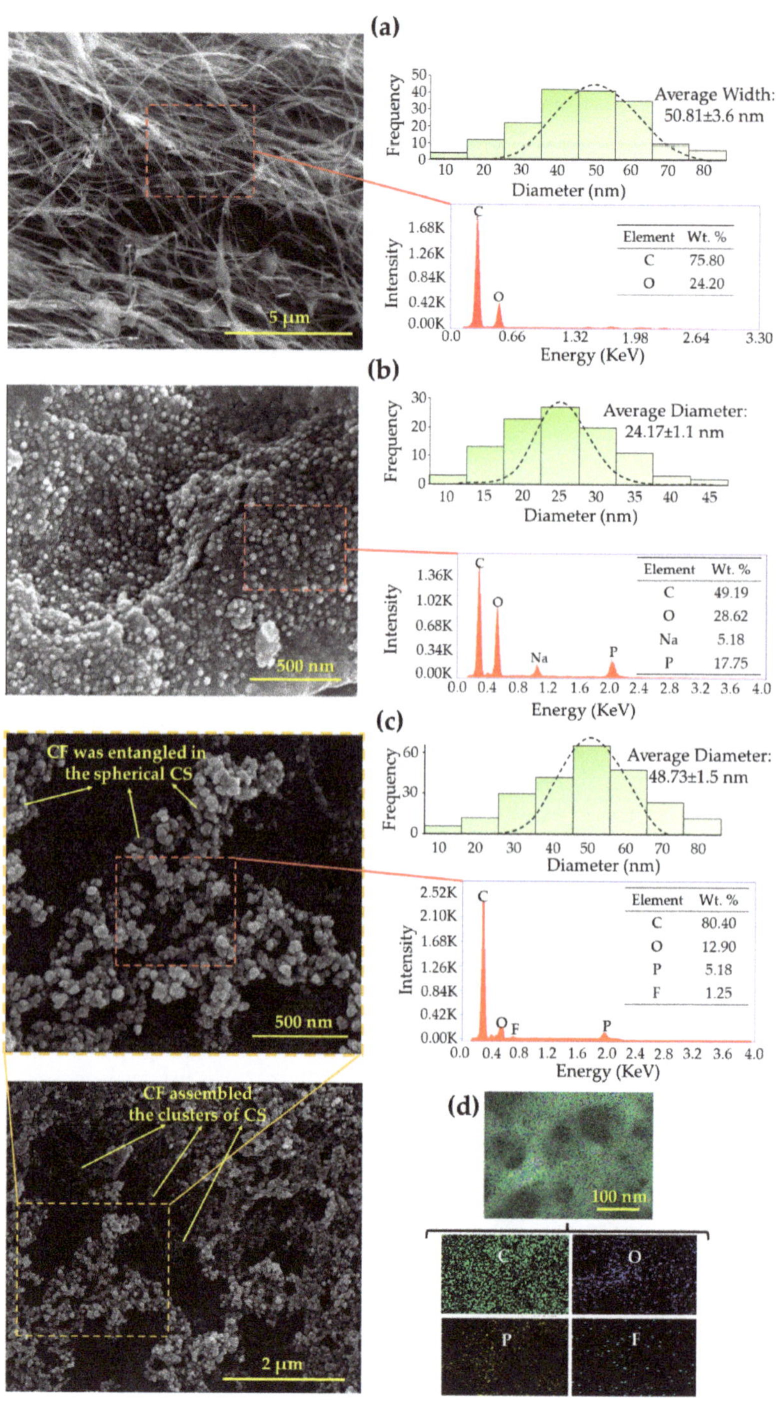

Figure 2. SEM image, the average diameter and EDX of (**a**) CF, (**b**) CS NPs and (**c**) CS-CF/5FU BNCs; (**d**) EDX layered image of CS-CF/5FU BNCs.

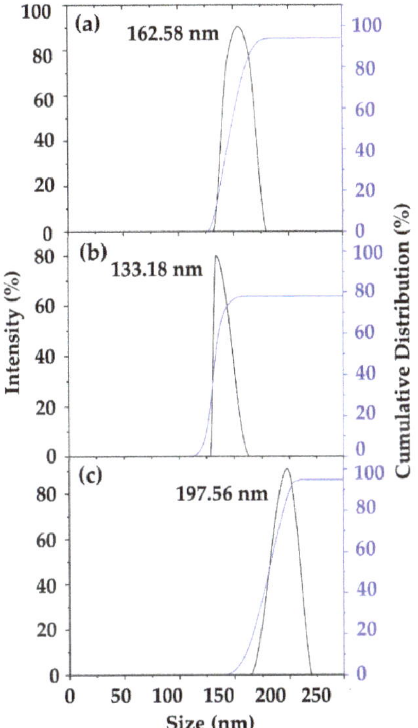

Figure 3. Hydrodynamic size distributions of (**a**) CF, (**b**) CS NPs and (**c**) CS-CF/5FU BNCs.

Table 1. Hydrodynamic size of the synthesized samples in the solutions with various pH values at 37 °C. Data is expressed as the mean ± standard deviation for triplicates within an individual experiment.

Sample	Hydrodynamic Particle Size (nm)		
	pH 1.2	**pH 7.4**	**pH 12**
CF	135.56 ± 2.87	174.43 ± 3.28	203.17 ± 4.81
CS	71.61 ± 5.54	140.09 ± 4.70	246.09 ± 5.10
CS-CF BNCs	109.03 ± 4.12	198 ± 3.25	275.34 ± 4.59
CS-CF/5FU BNCs	112.51 ± 4.09	203.52 ± 2.94	274.23 ± 5.11

3.4. Physicochemical Characterization of CS-CF/5FU BNCs Using Thermal Analysis

The thermal stability of the samples was determined by TGA. The results of TGA and differential thermogravimetric analysis (DTGA) of the fabricated samples are shown in Figure 4a–e. CF, CS NPs, CS-CF BNCs, CS-CF/5FU BNCs, and 5FU showed the main thermal degradation at 308.58, 236.54, 269.56, 298.10 and 284.60 °C with the final residue of 12.07, 14.53, 36.81, 32.50 and 0.91 weight % at 800 °C, respectively. During the thermal degradation, carbonyl and carboxyl groups could cause the reduction of the chain size and rapture the bonds of the polysaccharides of CS and CF [58]. It stated that the cellulosic materials may have a single-step degradation in nitrogen atmosphere, however, it is a two-step thermal degradation in air atmosphere [59]. In this manner, CF displayed a single-step degradation. CS-CF and CS-CF/5FU BNCs indicated a higher final residue % and thermal stability than CF and 5FU alone, showing the needs for the polymer blend of chitosan-cellulose to encapsulate anticancer drug in an advanced nano-carrier system.

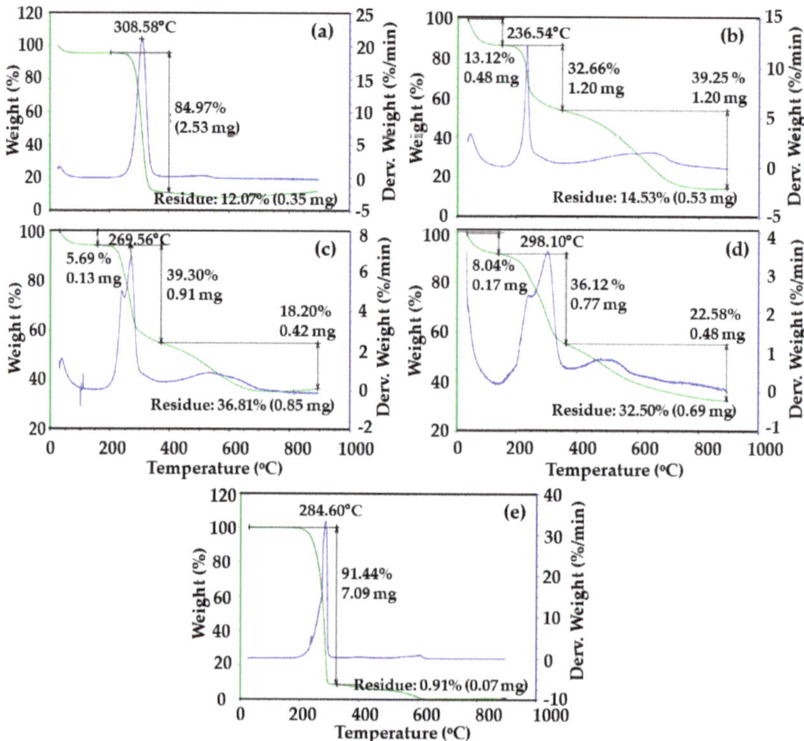

Figure 4. TGA and DTGA traces of (**a**) CF, (**b**) CS NPs, (**c**) CS-CF BNCs, (**d**) CS-CF/5FU BNCs and (**e**) 5FU.

So, for the percentage of the ingredient used for the preparation of CS-CF/5FU BNCs, it has been determined that approximately 64.91, 32.64, and 2.60 % of the constituent compounds are related to CS, CF and 5FU, respectively.

3.5. Physicochemical Characterization of CS-CF/5FU BNCs Using Fourier-Transform Infrared Spectroscopy

FTIR can identify changes in the chemical structure of the sample by generating an infrared absorption spectrum. The FTIR results of rice straw waste, CF, CS NPs, CS-CF BNCs, CS-CF/5FU BNCs and 5FU are indicated in Figure 5a–f, respectively. The bleaching procedure on rice straw waste potentially triggered the development of C-H aromatic hydrogen groups. From Figure 5b, the peaks at 1524 and 1750 cm^{-1} are for C–O–C bonds and bleaching of hemicellulose, respectively [41]. Both peaks at 760 and 491 cm^{-1} are attributed to elimination of silica (Si–O–Si stretching) [41]. Further results presented that the peaks at 3352, 2891 and 1100 cm^{-1} might determine the stretching vibrations of -OH groups, C-H stretching, and the cellulose network structure, respectively [3]. In the anomeric region (950–700 cm^{-1}), CF indicated a minor peak at 887 cm^{-1}, due to presence of the glycosidic –C$_1$ –O –C$_4$ deformation of the β-glycosidic bond [60]. This possibly indicated the efficient chemical treatments to isolate CF. The absorption peaks are different between CS NPs and CS-CF BNCs, showing contribution of OH or NH and changes in the sugar ring, Van der Waals forces, dipole moments and hydrogen bonds [61]. In CS-CF/5FU BNCs, the CO stretching vibration at 1654 cm^{-1} can be owed to a good interaction between the drug and its nano-carrier. Furthermore, an overlapping between OH and NH bonds of 5FU possibly demonstrated a band at 2750–3500 cm^{-1}, comparable with another study [19]. It stated that the amine groups of CS with carbonyl groups of CF may form functional groups of imines with carbon-nitrogen double bond [62].

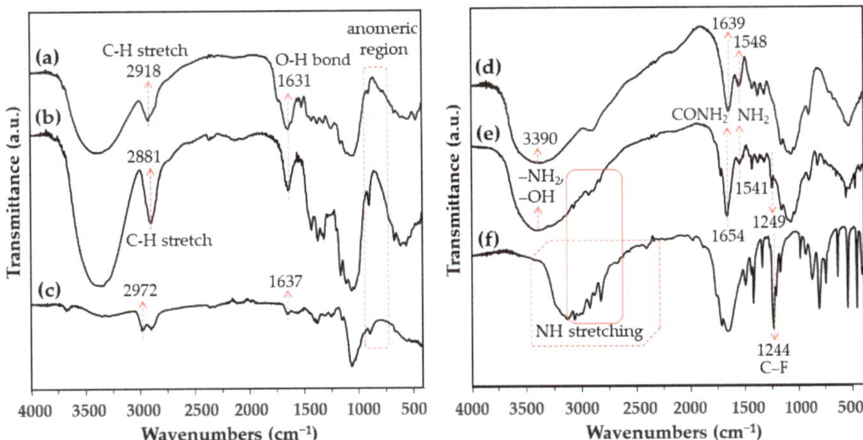

Figure 5. FTIR of (**a**) rice straw waste, (**b**) CF, (**c**) CS NPs, (**d**) CS-CF BNCs, (**e**) CS-CF/5FU BNCs and (**f**) 5FU.

As indicated in Figure 5d, CS-CF/5FU present a peak at 3385 cm^{-1}, because of –NH$_2$ and –OH stretching groups. The interaction between NH$_3^+$ groups of CS and phosphate groups of TPP led to presence of CONH$_2$ and NH$_2$ groups with peaks at 1639 and 1548 cm^{-1}, respectively [63]. Whereas, in the spectra of CS-CF/5FU, these peaks shifted to 1654 and 1541 cm^{-1}, respectively. The bands at 1411 and 1016 cm^{-1}, respectively, presented –CH$_2$ and P=O stretching vibrations from phosphate groups [63]. Furthermore, the primary amino group of CS reacted with the carbonyl groups of CF, whereas acetic acid dissolved CS to form hydrogen ion and affect the crosslinking structure from the Schiff base bond [64]. As stated that the primary amine group in CS could trigger the in situ gelation in the CS-based composites [65].

In addition, the development of the Schiff's base probably reduced the aldehyde groups on the CF surface for bonding with CS. From the spectrum of 5FU, the C–F stretching vibration is probably at 1244 cm^{-1}. This peak was shifted to 1249 cm^{-1} in the spectra of the CS-CF/5FU sample [19]. As reported [66], two protonated amide groups in the 5FU molecule may lead to a positive charge of 5FU and hydrogen bonds with carboxyl bonds, albeit, there were nearly ionized together to form a hydrogen bond. In turn, the 5FU molecules were possibly conjugated to the CS structure [66]. The FTIR spectrum of CS-CF/5FU BNCs indicated only a few peaks attributed to the drug, due to an efficient entrapping of the drug within the polysaccharide composites.

3.6. Swelling Analysis of CS-CF/5FU BNCs

The swelling analysis may determine the amount of the absorbed liquid by the polymer materials. The swelling properties of composites can act importantly in the drug loading and drug release performance. Figure 6a–d show the swelling kinetic of the fabricated samples in media at different pHs at 37 °C. The samples in the solutions at different pHs showed the main swelling ratio in the first 4 h. The increased swelling index ratio could be related to the natural property of the sample containing porous interconnected fibers [67]. The sample approximately showed its equilibrium state after 12 h, comparable with different studies [32,67]. With an increasing time and pH value, the swelling increased that CF, CS NPs, CS-CF BNCs and CS-CF/5FU BNCs indicated the maximum swelling ratio of 2.91 ± 0.11, 3.49 ± 0.09, 3.77 ± 0.07 and 3.85 ± 0.10 in media at pH 12, respectively, after 36 h. Similarly, in different studies, the pH of solution controlled the swelling performances of polymer complexes, including polyvinyl alcohol/chitosan/TiO$_2$ nanofibers [68], chitosan cross-linked poly (acrylic acid) hydrogels [69] and hydrogel composites of carboxymethylstarch-g-poly (acrylic acid)/palygorskite/starch/sodium alginate [70]. As presented in Figure 6a–d, CS NPs and the CS-CF composites displayed higher swelling properties compared to CF. Therefore, the CS structure domain the swelling properties of the CS-CF composites. The increasing swelling properties of the tested samples were in line with the hydrodynamic size in the solution with various pH values. The above swelling results could indicate the necessity of using CS to coat the CF matrix for increasing pH sensitivity of the CS-CF composites.

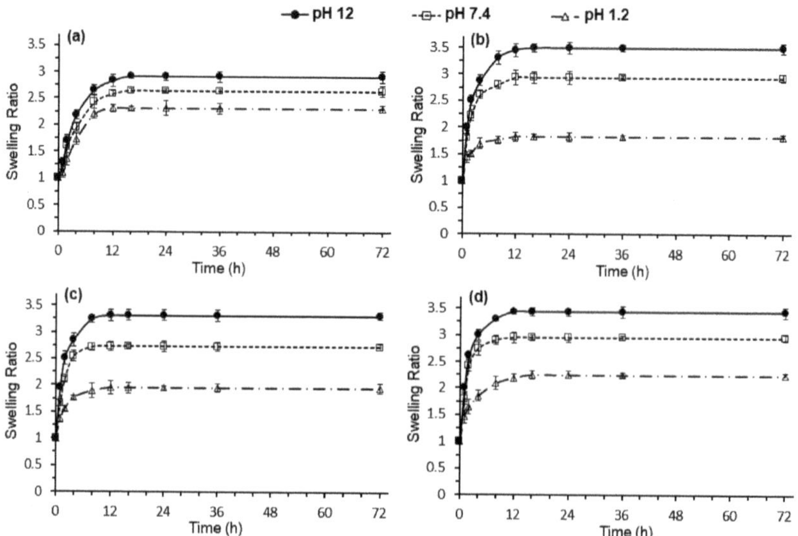

Figure 6. Swelling kinetic of (a) CF, (b) CS NPs, (c) CS-CF BNCs, and (d) CS-CF/5FU BNCs initiated from the wet state in three different media at pH 1.2, 7.4, and 12. Data is expressed as the mean ± standard deviation for triplicates within an individual experiment.

3.7. Drug Loading and Encapsulation Efficiency Percentage of CS-CF/5FU BNCs

Encapsulation efficiency was used to estimate the successful drug loading ratio in a nano-carrier system. Figure 7a(i–iii) indicate the 5FU absorbance, calibration curve at λ_{max} = 266 nm from known concentrations of 5FU (5–25 ppm), and UV-vis spectra of the unentrapped 5FU, respectively. From the UV absorbance and Equation (1) the encapsulation efficiency value was estimated to be 86 ± 2.75%. The alkali treated CF possibly possessed open bonds and good swelling properties to host 5FU [71]. The CS-CF composites could show the OH groups for effectively binding and releasing 5FU by ionic interactions, as reported [72]. Furthermore, 5FU is a heterocyclic aromatic organic compound with a low molecular weight to be diffused well within open pores and substrate of the CF [73]; whereas the CS coated both CF and 5FU to improve encapsulation efficiency. From the value of encapsulation efficiency, the synthesized composites could be a suitable nano-carrier for the 5FU encapsulation in cancer treatment.

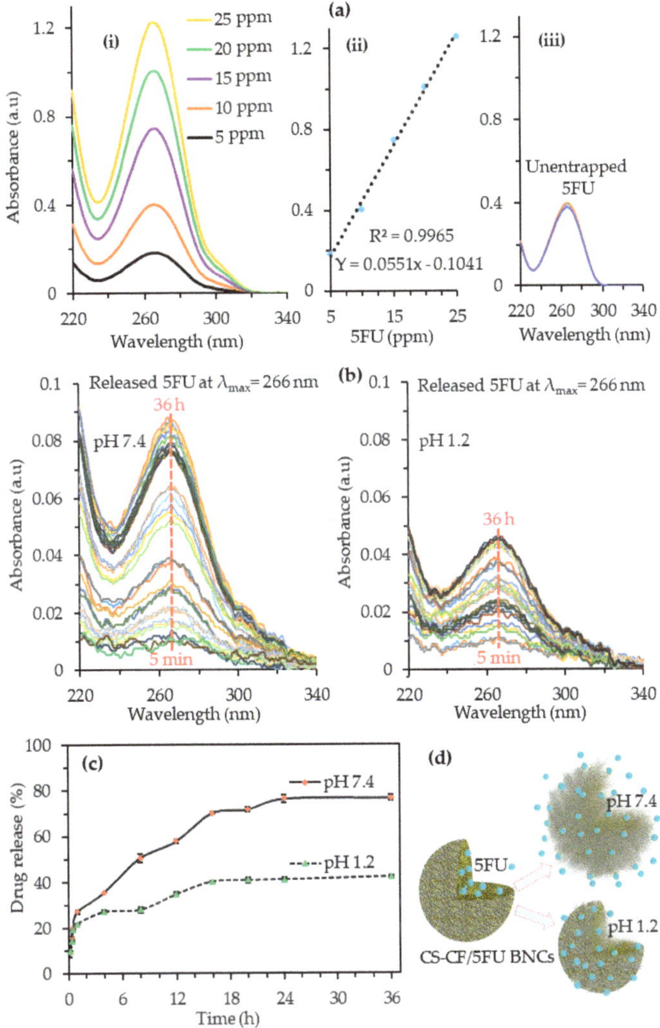

Figure 7. (**a**) (**i**) The 5FU absorbance and (**ii**) calibration curve at λ_{max} = 266 nm from known concentrations of 5FU (5–25 ppm); (**iii**) UV-vis spectra of the unentrapped 5FU; (**b**) UV-vis spectra and (**c**) the release kinetics of 5FU from CS-CF/5FU BNCs; (**d**) a schematic of possible drug release. Data is expressed as the mean ± standard deviation for triplicates within an individual experiment.

3.8. In Vitro Drug Release of CS-CF/5FU BNCs

In vitro drug release analysis was used to determine the release kinetics of the drug-loaded samples. Figure 7b,c shows UV-vis spectra and the release kinetics of 5FU from CS-CF/5FU BNCs, respectively. As seen in Figure 7c, the time taken was 8 h for the drug release of 50.60 ± 1.88% and 28.04 ± 1.14% at the release fluids with pH 7.4 and pH 1.2, respectively. After 36 h, the total 5FU release was lower in the simulated acidic fluid at pH 1.2 (42.37 ± 0.43%) compared to that in the simulated colorectal fluid at pH 7.4 (76.82 ± 1.29%). This can be explained that CS-CF/5FU BNCs became smaller in the acidic media [67]. In the human body, the above may possibly result in a safe transferring of CS-CF/5FU BNCs through the acidic fluid of the stomach and then it may begin to break up in the colorectal fluid [67]. The low molecular weight of 5FU may cause a tendency for consecutive release, albeit, the cellulosic material as an amphipathic drug carrier and even excipient may modulate the release dosage [73,74]. The drug release showed an analogous trend with the swelling ratio of the samples, as schematically depicted in Figure 7d. An investigation stated that the controlled release ratio could be related to the interaction between 5FU and –COOH functionalities presented in the cellulosic material as it decreased when the swelling capacity increased [19]. The swelling may lead to polymer relaxation and gelling of 5FU to obtain a controlled drug release profile from CS-CF/5FU BNCs [74]. CS-CF/5FU BNCs presented a sustained and pH-sensitive release behavior, therefore, it may be considered as an attractive drug formulation candidate for modern CRC treatments.

3.9. In Vitro Cytotoxicity Assays

The biocompatibility and potential anticancer effects of the samples were studied by in vitro cytotoxicity assays and antibacterial assay (Supplementary Materials Figure S1). The cytotoxicity of CF, CS NPs, CS-CF BNCs, CS-CF/5FU BNCs against HCT116 CRC and CCD112 normal colon cell lines are shown in Figure 8a–d. CF exhibited almost no prominent killing towards both normal and cancer cell lines (Figure 8a). Different reports stated that the wood-based nanocelluloses possess no adverse toxicity and based on ISO standard 10993-5, it does not have toxicity above 30% [75]. From Figure 8b, 15.62 µg/mL of CS NPs was enough to eliminate $26.46 \pm 6.52\%$ of the CRC cells without any toxicity on the CCD112 normal cell. In addition, the highest concentration (500 µg/mL) of CF, CS NPs, and CS-CF BNCs showed an appropriate biocompatibility without any damage against the normal cells. Among the samples without the loaded-5FU, CS-CF BNCs displayed better anticancer effect ($41.62 \pm 3.15\%$), showing a suitable contribution between CS and CF in the composites. As previously stated [15], the therapeutic potency of the chitosan-based samples could be explained that the positive charge of chitosan neutralized the negative charge on the tumor cell surface [14]. A different study also reported that the strong bonding between the primary amino group of chitosan and the carbonyl of cellulose could block bacteria from crossing the biocomposites [49]. It was indicated that the anticancer effects and permeation enhancement of CS-based materials could be related to the primary amine group in CS [65].

It can be noticed from Figure 8d that certain concentrations of CS-CF/5FU BNCs gave only negligible damage on the normal cells, while exhibiting effective anti-proliferation actions towards the cancer cells. Notably, CS-CF/5FU BNCs at 250 µg/mL concentration killed $56.42 \pm 0.41\%$ of the cancer cells and just $8.16 \pm 2.11\%$ of the normal cells. This anticancer effect might be attributed to the properties of the synthesized double polysaccharides to encapsulate a sufficient amount of 5FU for potentially improving mobility, binding ability, and colloidal stability to attach on the cell surface, as reported earlier [75]. Besides, based on a different study [17], polysaccharide complex might conjugate onto cancer cells to improve selectivity and decrease the drug leakage from the selected cells for saving the drug from unwanted degradation and elimination [16]. The polymeric network of CS-CF entrapped 5FU to potentially obtain a prolonged release dosage of the drug, similar to a separate report [16]. In the drug-loaded chitosan materials, the primary amine groups can possibly cause the controlled drug release [65]. As shown in Table 2, no activity was seen in the synthesized samples on the normal cells which suggests its high biocompatibility. Among all the synthesized samples, the IC_{50} happened only for the drug loaded sample that CS-CF/5FU BNCs showed the IC_{50} of 228.27 µg/mL on the cancer cells. It could be understood from the results of in vitro cytotoxicity assays that the polysaccharide samples of CF and CS have a suitable biocompatibility and CS-CF/5FU BNC suggests promising anticancer actions.

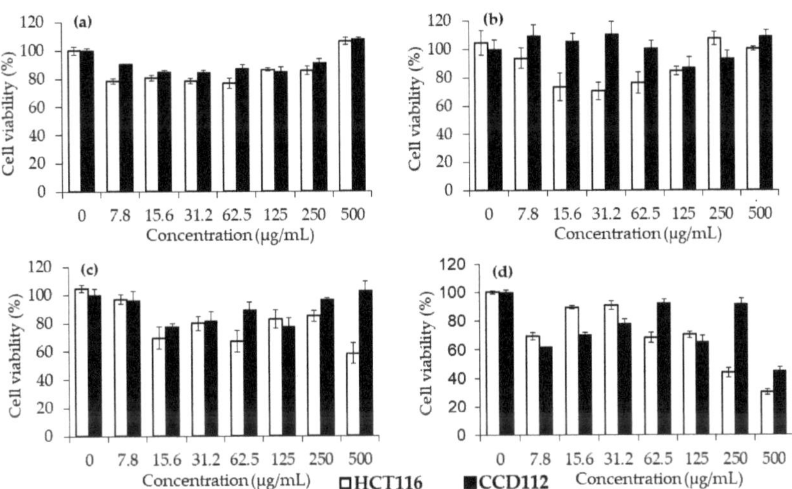

Figure 8. In vitro anticancer activity of (**a**) CF, (**b**) CS NPs, (**c**) CS-CF BNCs and (**d**) CS-CF/5FU BNCs against CCD112 colon normal and HCT116 CRC cell lines. Three independent experiments were performed and the data are expressed as the mean ± standard deviation for all triplicates within an individual experiment.

Table 2. IC_{50} of the synthesized samples.

Sample	IC_{50} (µg/mL)	
	HCT-116	**CCD 112**
CF	>500	>500
CS NPs,	>500	>500
CS-CF BNCs	>500	>500
CS-CF/5FU BNCs	228.27	>500

4. Conclusions

In this study, lignin and hemicellulose of rice straw waste were degraded by a series of chemical modifications include dewaxing, bleaching, and alkali treatments to liberate CF containing organized nanofibrils. By using the ionic gelation method, CS-CF composites encapsulated 5FU to synthesize nanodrug candidate of CS-CF/5FU BNCs for in vitro CRC analysis. The composites indicated the XRD and FTIR peaks related to the CS cross-linked CF. The SEM images of CS-CF/5FU BNCs showed that the spherical CS NPs covered the rod-shaped CF reinforcement. Based on swelling and hydrodynamic size analysis, the composites indicated a pH responsive behavior. Analogously, CS-CF/5FU BNCs exhibited a lower release ratio in media simulating the gastric fluid at pH 1.2 compared to the colorectal fluid at pH 7.4. In cytotoxicity assays, all the synthesized polysaccharide samples did not indicate the IC_{50} against the CCD112 normal colon cells, showing their proper biocompatibility and safety for anticancer applications. Among the fabricated samples, CS-CF/5FU BNCs only displayed the IC_{50} value (228.27 µg/mL) against the HCT116 cancer cells. Further results presented that the 250 µg/mL concentration of CS-CF/5FU BNCs showed better anticancer actions than other concentrations. The intriguing implication of this research is that using a very low dosage of 5-FU encapsulated in CS/CF composites could reduce the manufacturing cost and make it an effective green nanodrug candidate for anticancer actions against CRC.

Supplementary Materials: The following are available online at https://www.mdpi.com/article/10.3390/nano11071691/s1, Figure S1: Antibacterial activity of CF, CS NPs, CS-CF BNCs and CS-CF/5-FU BNCs against (a) gram-positive *Staphylococcus aureus* and (b) gram-negative *Escherichia coli*.

Author Contributions: The manuscript was written through contributions of all authors. Conceptualization, methodology, writing—original draft, fabrication of the samples, the physiochemical analysis, drug loading and drug release study, contribution for in vitro anticancer assays, M.Y.; review & editing, analysis of anticancer assays, H.-Y.C.; supervision, validation and writing—review & editing, analysis of anticancer assays, S.-Y.T.; contribution in SEM analysis, drug loading, and drug release study, P.K.; investigation, in vitro anticancer assays, M.L.-K.S.; supervision, validation, N.A.B.C.S.; supervision, project administration, formal analysis, sample preparation, validation and writing—review & editing, funding acquisition, K.S. All authors have read and agreed to the published version of the manuscript.

Funding: This research was funded by Takasago Thermal Engineering Co. Ltd. grants including R.K130000.7343.4B422, 4B472 and 4B631. Also, we have express gratitude to the research management center (RMC) of Universiti Teknologi Malaysia (UTM) and Malaysia-Japan International Institute of Technology (MJIIT).

Institutional Review Board Statement: Not applicable.

Informed Consent Statement: Not applicable.

Data Availability Statement: The data presented in this study are available on request from the corresponding author.

Acknowledgments: Special thanks to School of Medical and Life Sciences, Sunway University Malaysia for the cell culture and laboratory facilities.

Conflicts of Interest: The authors declare no conflict of interest. The funders had no role in the design of the study; in the collection, analyses, or interpretation of data; in the writing of the manuscript, or in the decision to publish the results.

References

1. Yusefi, M.; Yee, O.S.; Shameli, K. Bio-mediated production and characterisation of magnetic nanoparticles using fruit peel extract. *J. Res. Nanosci. Nanotechnol.* **2021**, *1*, 53–61. [CrossRef]
2. Perumal, A.B.; Sellamuthu, P.S.; Nambiar, R.B.; Sadiku, E.R. Development of polyvinyl alcohol/chitosan bio-nanocomposite films reinforced with cellulose nanocrystals isolated from rice straw. *Appl. Surf. Sci.* **2018**, *449*, 591–602. [CrossRef]
3. Chen, X.; Yu, J.; Zhang, Z.; Lu, C. Study on structure and thermal stability properties of cellulose fibers from rice straw. *Carbohydr. Polym.* **2011**, *85*, 245–250. [CrossRef]
4. Ahmad, R.; Deng, Y.; Singh, R.; Hussain, M.; Shah, M.A.A.; Elingarami, S.; He, N.; Sun, Y. Cutting edge protein and carbohydrate-based materials for anticancer drug delivery. *J. Biomed. Nanotechnol.* **2018**, *14*, 20–43. [CrossRef] [PubMed]
5. Aimonen, K.; Suhonen, S.; Hartikainen, M.; Lopes, V.R.; Norppa, H.; Ferraz, N.; Catalán, J. Role of surface chemistry in the in vitro lung response to nanofibrillated cellulose. *Nanomaterials* **2021**, *11*, 389. [CrossRef]
6. Kumar, M.N.R. A review of chitin and chitosan applications. *React. Funct. Polym.* **2000**, *46*, 1–27. [CrossRef]
7. Liu, S.; Zhang, J.; Cui, X.; Guo, Y.; Zhang, X.; Hongyan, W. Synthesis of chitosan-based nanohydrogels for loading and release of 5-fluorouracil. *Colloids Surf. A Phys. Eng. Asp.* **2016**, *490*, 91–97. [CrossRef]
8. Ruman, U.; Buskaran, K.; Pastorin, G.; Masarudin, M.J.; Fakurazi, S.; Hussein, M.Z. Synthesis and characterization of chitosan-based nanodelivery systems to enhance the anticancer effect of sorafenib drug in hepatocellular carcinoma and colorectal adenocarcinoma cells. *Nanomaterials* **2021**, *11*, 497. [CrossRef]
9. Berger, J.; Reist, M.; Mayer, J.M.; Felt, O.; Peppas, N.; Gurny, R. Structure and interactions in covalently and ionically crosslinked chitosan hydrogels for biomedical applications. *Eur. J. Pharm. Biopharm.* **2004**, *57*, 19–34. [CrossRef]
10. Schwaiger, D.; Lohstroh, W.; Müller-Buschbaum, P. Investigation of molecular dynamics of a PTB7: PCBM polymer blend with quasi-elastic neutron scattering. *ACS Appl. Polym. Mater.* **2020**, *2*, 3797–3804. [CrossRef]
11. Ogueri, K.S.; Ogueri, K.S.; Allcock, H.R.; Laurencin, C.T. A regenerative polymer blend composed of glycylglycine ethyl ester-substituted polyphosphazene and poly (lactic-co-glycolic acid). *ACS Appl. Polym. Mater.* **2020**, *2*, 1169–1179. [CrossRef] [PubMed]
12. Ignacz, G.; Fei, F.; Szekely, G. Ion-stabilized membranes for demanding environments fabricated from polybenzimidazole and its blends with polymers of intrinsic microporosity. *ACS Appl. Polym. Mater.* **2018**, *1*, 6349–6356. [CrossRef]
13. Chatterjee, S.; Hui, P.C.-L.; Kan, C.-W.; Wang, W. Dual-responsive (pH/temperature) Pluronic F-127 hydrogel drug delivery system for textile-based transdermal therapy. *Sci. Rep.* **2019**, *9*, 1–13. [CrossRef] [PubMed]
14. Anirudhan, T.S.; Christa, J. Multi-polysaccharide based stimuli responsive polymeric network for the in vitro release of 5-fluorouracil and levamisole hydrochloride. *New J. Chem.* **2017**, *41*, 11979–11990. [CrossRef]
15. Mohammed, M.A.; Syeda, J.; Wasan, K.M.; Wasan, E.K. An overview of chitosan nanoparticles and its application in non-parenteral drug delivery. *Pharmaceutics* **2017**, *9*, 53. [CrossRef]

16. Banerjee, A.; Pathak, S.; Subramanium, V.D.; Dharanivasan, G.; Murugesan, R.; Verma, R.S. Strategies for targeted drug delivery in treatment of colon cancer: Current trends and future perspectives. *Drug Discov. Today* **2017**, *22*, 1224–1232. [CrossRef]
17. Lokich, J. Infusional 5-FU: Historical evolution, rationale, and clinical experience. *Oncology* **1998**, *12*, 19–22.
18. Anirudhan, T.S.; Binusreejayan, R.S.R. Synthesis and characterization of chitosan based multilayer and pH sensitive co-polymeric system for the targeted delivery of 5-fluorouracil, an in vitro study. *Int. J. Polym. Mater* **2014**, *63*, 539–548. [CrossRef]
19. Anirudhan, T.S.; Nima, J.; Divya, P.L. Synthesis, characterization and in vitro cytotoxicity analysis of a novel cellulose based drug carrier for the controlled delivery of 5-fluorouracil, an anticancer drug. *Appl. Surf. Sci.* **2015**, *355*, 64–73. [CrossRef]
20. Yusefi, M.; Shameli, K.; Jahangirian, H.; Teow, S.-Y.; Umakoshi, H.; Saleh, B.; Rafiee-Moghaddam, R.; Webster, T.J. The potential anticancer activity of 5-fluorouracil loaded in cellulose fibers isolated from rice straw. *Int. J. Nanomed.* **2020**, *15*, 5417–5432. [CrossRef]
21. Chatterjee, S.; Hui, P.C.-L.; Kan, C.-W. Thermoresponsive hydrogels and their biomedical applications: Special insight into their applications in textile based transdermal therapy. *Polymers* **2018**, *10*, 480. [CrossRef] [PubMed]
22. Yusefi, M.; Shameli, K.; Jumaat, A.F. Preparation and properties of magnetic iron oxide nanoparticles for biomedical applications: A brief review. *J. Adv. Res. Mater. Sci.* **2020**, *75*, 10–18. [CrossRef]
23. Yusefi, M.; Shameli, K. Nanocellulose as a Vehicle for Drug Delivery and Efficiency of Anticancer Activity: A Short-Review. *J. Res. Nanosci. Nanotechnol.* **2021**, *1*, 30–43. [CrossRef]
24. Topuz, F.; Kilic, M.E.; Durgun, E.; Szekely, G. Fast-dissolving antibacterial nanofibers of cyclodextrin/antibiotic inclusion complexes for oral drug delivery. *J. Colloid Interface Sci.* **2021**, *585*, 184–194. [CrossRef]
25. Shehabeldine, A.; El-Hamshary, H.; Hasanin, M.; El-Faham, A.; Al-Sahly, M. Enhancing the antifungal activity of griseofulvin by incorporation a green biopolymer-based nanocomposite. *Polymers* **2021**, *13*, 542. [CrossRef]
26. Bae, J.; Park, J.W. Preparation of an injectable depot system for long-term delivery of alendronate and evaluation of its anti-osteoporotic effect in an ovariectomized rat model. *Int. J. Pharm.* **2015**, *480*, 37–47. [CrossRef]
27. Figueiredo, P.; Lintinen, K.; Kiriazis, A.; Hynninen, V.; Liu, Z.; Bauleth-Ramos, T.; Rahikkala, A.; Correia, A.; Kohout, T.; Sarmento, B. In vitro evaluation of biodegradable lignin-based nanoparticles for drug delivery and enhanced antiproliferation effect in cancer cells. *Biomaterials* **2017**, *121*, 97–108. [CrossRef]
28. Ntoutoume, G.M.N.; Granet, R.; Mbakidi, J.P.; Brégier, F.; Léger, D.Y.; Fidanzi-Dugas, C.; Lequart, V.; Joly, N.; Liagre, B.; Chaleix, V. Development of curcumin–cyclodextrin/cellulose nanocrystals complexes: New anticancer drug delivery systems. *Bioorganic Med. Chem. Lett.* **2016**, *26*, 941–945. [CrossRef]
29. Li, J.; Wang, Y.; Zhang, L.; Xu, Z.; Dai, H.; Wu, W. Nanocellulose/gelatin composite cryogels for controlled drug release. *Chem. Eng.* **2019**, *7*, 6381–6389. [CrossRef]
30. Mohan, D.; Khairullah, N.F.; How, Y.P.; Sajab, M.S.; Kaco, H. 3D printed laminated CaCO3-nanocellulose films as controlled-release 5-fluorouracil. *Polymers* **2020**, *12*, 986. [CrossRef]
31. Latifi, N.; Asgari, M.; Vali, H.; Mongeau, L. A tissue-mimetic nano-fibrillar hybrid injectable hydrogel for potential soft tissue engineering applications. *Sci. Rep.* **2018**, *8*, 1–18. [CrossRef]
32. Bullo, S.; Buskaran, K.; Baby, R.; Dorniani, D.; Fakurazi, S.; Hussein, M.Z. Dual drugs anticancer nanoformulation using graphene oxide-PEG as nanocarrier for protocatechuic acid and chlorogenic acid. *Pharm. Res.* **2019**, *36*, 91. [CrossRef] [PubMed]
33. Yew, Y.P.; Shameli, K.; Mohamad, S.E.; Lee, K.X.; Teow, S.-Y. Green synthesized montmorillonite/carrageenan/Fe$_3$O$_4$ nanocomposites for ph-responsive release of protocatechuic acid and its anticancer activity. *Int. J. Mol. Sci.* **2020**, *21*, 4851. [CrossRef] [PubMed]
34. Yusefi, M.; Shameli, K.; Ali, R.R.; Pang, S.-W.; Teow, S.-Y. Evaluating anticancer activity of plant-mediated synthesized iron oxide nanoparticles using Punica Granatum fruit peel extract. *J. Mol. Struct.* **2020**, *1204*, 127539. [CrossRef]
35. Yusefi, M.; Shameli, K.; Yee, O.S.; Teow, S.-Y.; Hedayatnasab, Z.; Jahangirian, H.; Webster, T.J.; Kuča, K. Green synthesis of Fe$_3$O$_4$ nanoparticles stabilized by a Garcinia mangostana fruit peel extract for hyperthermia and anticancer activities. *Int. J. Nanomed.* **2021**, *16*, 2515. [CrossRef]
36. Sukri, S.N.A.M.; Shameli, K.; Wong, M.M.-T.; Teow, S.-Y.; Chew, J.; Ismail, N.A. Cytotoxicity and antibacterial activities of plant-mediated synthesized zinc oxide (ZnO) nanoparticles using Punica granatum (pomegranate) fruit peels extract. *J. Mol. Struct.* **2019**, *1189*, 57–65. [CrossRef]
37. Yusefi, M.; Shameli, K.; Hedayatnasab, Z.; Teow, S.-Y.; Ismail, U.N.; Azlan, C.A.; Ali, R.R. Green synthesis of Fe$_3$O$_4$ nanoparticles for hyperthermia, magnetic resonance imaging and 5-fluorouracil carrier in potential colorectal cancer treatment. *Res. Chem. Intermed.* **2021**, 1–20. [CrossRef]
38. Abe, K.; Yano, H. Comparison of the characteristics of cellulose microfibril aggregates of wood, rice straw and potato tuber. *Cellulose* **2009**, *16*, 1017–1023. [CrossRef]
39. Srasri, K.; Thongroj, M.; Chaijiraaree, P.; Thiangtham, S.; Manuspiya, H.; Pisitsak, P.; Ummartyotin, S. Recovery potential of cellulose fiber from newspaper waste: An approach on magnetic cellulose aerogel for dye adsorption material. *Int. J. Biol. Macromol.* **2018**, *119*, 662–668. [CrossRef]
40. Lu, P.; Hsieh, Y.-L. Preparation and characterization of cellulose nanocrystals from rice straw. *Carbohydr. Polym.* **2012**, *87*, 564–573. [CrossRef]
41. Cai, M.; Takagi, H.; Nakagaito, A.N.; Katoh, M.; Ueki, T.; Waterhouse, G.I.; Li, Y. Influence of alkali treatment on internal microstructure and tensile properties of abaca fibers. *Ind. Crop. Prod.* **2015**, *65*, 27–35. [CrossRef]

42. Kunjachan, S.; Jose, S.; Lammers, T. Understanding the mechanism of ionic gelation for synthesis of chitosan nanoparticles using qualitative techniques. *Asian J. Pharm.* **2014**, *4*. [CrossRef]

43. Hps, A.K.; Saurabh, C.K.; Adnan, A.; Fazita, M.N.; Syakir, M.; Davoudpour, Y.; Rafatullah, M.; Abdullah, C.; Haafiz, M.; Dungani, R. A review on chitosan-cellulose blends and nanocellulose reinforced chitosan biocomposites: Properties and their applications. *Carbohydr. Polym.* **2016**, *150*, 216–226. [CrossRef]

44. Bozoğlan, B.K.; Duman, O.; Tunç, S. Preparation and characterization of thermosensitive chitosan/carboxymethylcellulose/ scleroglucan nanocomposite hydrogels. *Int. J. Biol. Macromol.* **2020**, *162*, 781–797. [CrossRef] [PubMed]

45. Yang, J.; Duan, J.; Zhang, L.; Lindman, B.; Edlund, H.; Norgren, M. Spherical nanocomposite particles prepared from mixed cellulose–chitosan solutions. *Cellulose* **2016**, *23*, 3105–3115. [CrossRef]

46. Samy, M.; Abd El-Alim, S.H.; Amin, A.; Ayoub, M.M. Formulation, characterization and in vitro release study of 5-fluorouracil loaded chitosan nanoparticles. *Int. J. Biol. Macromol.* **2020**, *156*, 783–791. [CrossRef]

47. Nguyen, D.H. Potential 5-fluorouracil encapsulated mPEG-Chitosan nanogels for controlling drug release. *JAMPS* **2017**, 1–7. [CrossRef]

48. Hosokawa, J.; Nishiyama, M.; Yoshihara, K.; Kubo, T.; Terabe, A. Reaction between chitosan and cellulose on biodegradable composite film formation. *Ind. Eng. Chem. Res.* **1991**, *30*, 788–792. [CrossRef]

49. Jiang, F.; Han, S.; Hsieh, Y.-L. Controlled defibrillation of rice straw cellulose and self-assembly of cellulose nanofibrils into highly crystalline fibrous materials. *RSC Adv.* **2013**, *3*, 12366–12375. [CrossRef]

50. Ali, M.E.A.; Aboelfadl, M.M.S.; Selim, A.M.; Khalil, H.F.; Elkady, G.M. Chitosan nanoparticles extracted from shrimp shells, application for removal of Fe (II) and Mn (II) from aqueous phases. *Sep. Sci. Technol.* **2018**, *53*, 2870–2881. [CrossRef]

51. Nivethaa, E.; Dhanavel, S.; Narayanan, V.; Vasu, C.A.; Stephen, A. An in vitro cytotoxicity study of 5-fluorouracil encapsulated chitosan/gold nanocomposites towards MCF-7 cells. *RSC Adv.* **2015**, *5*, 1024–1032. [CrossRef]

52. Guo, J.; Wang, Y.; Wang, J.; Zheng, X.; Chang, D.; Wang, S.; Jiang, T. A novel nanogel delivery of poly-α, β-polyasparthydrazide by reverse microemulsion and its redox-responsive release of 5-Fluorouridine. *Asian J. Pharm. Sci.* **2016**, *11*, 735–743. [CrossRef]

53. Zhu, W.; Wan, L.; Zhang, C.; Gao, Y.; Zheng, X.; Jiang, T.; Wang, S. Exploitation of 3D face-centered cubic mesoporous silica as a carrier for a poorly water soluble drug: Influence of pore size on release rate. *Mater. Sci. Eng. C* **2014**, *34*, 78–85. [CrossRef]

54. Maluin, F.N.; Hussein, M.Z.; Yusof, N.A.; Fakurazi, S.; Idris, A.S.; Zainol Hilmi, N.H.; Jeffery Daim, L.D. Preparation of chitosan–hexaconazole nanoparticles as fungicide nanodelivery system for combating Ganoderma disease in oil palm. *Molecules* **2019**, *24*, 2498. [CrossRef]

55. De Mesquita, J.P.; Donnici, C.L.; Teixeira, I.F.; Pereira, F.V. Bio-based nanocomposites obtained through covalent linkage between chitosan and cellulose nanocrystals. *Carbohydr. Polym.* **2012**, *90*, 210–217. [CrossRef] [PubMed]

56. Hussain, Z.; Katas, H.; Amin, M.C.I.M.; Kumolosasi, E.; Buang, F.; Sahudin, S. Self-assembled polymeric nanoparticles for percutaneous co-delivery of hydrocortisone/hydroxytyrosol: An ex vivo and in vivo study using an NC/Nga mouse model. *Int. J. Pharm.* **2013**, *444*, 109–119. [CrossRef] [PubMed]

57. Pech-Cohuo, S.-C.; Canche-Escamilla, G.; Valadez-González, A.; Fernández-Escamilla, V.V.A.; Uribe-Calderon, J. Production and modification of cellulose nanocrystals from Agave tequilana weber waste and its effect on the melt rheology of PLA. *Int. J. Polym. Sci.* **2018**, *2018*. [CrossRef]

58. Santmartí, A.; Lee, K.-Y. Crystallinity and Thermal Stability of Nanocellulose. In *Nanocellulose Sustainability. Production, Properties, Applications and Case Studies*; CRC Press: Boca Raton, FL, USA, 2018; pp. 67–86.

59. Jiang, F.; Hsieh, Y.-L. Chemically and mechanically isolated nanocellulose and their self-assembled structures. *Carbohydr. Polym.* **2013**, *95*, 32–40. [CrossRef] [PubMed]

60. Kavaz, D.; Kirac, F.; Kirac, M.; Vaseashta, A. Low releasing mitomycin c molecule encapsulated with chitosan nanoparticles for intravesical installation. *J. Biomater. Nanobiotechnol.* **2017**, *8*, 203–219. [CrossRef]

61. Area, M.C.; Ceradame, H. Paper aging and degradation: Recent findings and research methods. *Bioresources* **2011**, *6*, 5307–5337.

62. Lustriane, C.; Dwivany, F.M.; Suendo, V.; Reza, M. Effect of chitosan and chitosan-nanoparticles on post harvest quality of banana fruits. *J. Plant Biotechnol.* **2018**, *45*, 36–44. [CrossRef]

63. Puchtler, H.; Meloan, S. On Schiff's bases and aldehyde-Fuchsin: A review from H. Schiff to RD Lillie. *Histochemistry* **1981**, *72*, 321–332. [CrossRef] [PubMed]

64. Ali, A.; Ahmed, S. A review on chitosan and its nanocomposites in drug delivery. *Int. J. Biol. Macromol.* **2018**, *109*, 273–286. [CrossRef]

65. Zheng, X.-F.; Lian, Q.; Yang, H.; Wang, X. Surface molecularly imprinted polymer of chitosan grafted poly (methyl methacrylate) for 5-fluorouracil and controlled release. *Sci. Rep.* **2016**, *6*, 21409. [CrossRef]

66. Bhandari, J.; Mishra, H.; Mishra, P.K.; Wimmer, R.; Ahmad, F.J.; Talegaonkar, S. Cellulose nanofiber aerogel as a promising biomaterial for customized oral drug delivery. *Int. J. Nanomed.* **2017**, *12*, 2021. [CrossRef] [PubMed]

67. Nugraheni, A.D.; Purnawati, D.; Rohmatillah, A.; Mahardika, D.N.; Kusumaatmaja, A. Swelling of PVA/chitosan/TiO$_2$ nanofibers membrane in different pH. *Mater. Sci. Forum.* **2020**, *990*, 220–224. [CrossRef]

68. Wang, Y.; Wang, J.; Yuan, Z.; Han, H.; Li, T.; Li, L.; Guo, X. Chitosan cross-linked poly (acrylic acid) hydrogels: Drug release control and mechanism. *Colloids Surf. B* **2017**, *152*, 252–259. [CrossRef] [PubMed]

69. Gao, J.; Fan, D.; Song, P.; Zhang, S.; Liu, X. Preparation and application of pH-responsive composite hydrogel beads as potential delivery carrier candidates for controlled release of berberine hydrochloride. *R. Soc. Open Sci.* **2020**, *7*, 200676. [CrossRef]

70. Kadry, G. Comparison between gelatin/carboxymethyl cellulose and gelatin/carboxymethyl nanocellulose in tramadol drug loaded capsule. *Heliyon* **2019**, *5*, e02404. [CrossRef]
71. Seabra, A.B.; Bernardes, J.S.; Fávaro, W.J.; Paula, A.J.; Durán, N. Cellulose nanocrystals as carriers in medicine and their toxicities: A review. *Carbohyd. Polym.* **2018**, *181*, 514–527. [CrossRef]
72. Illangakoon, U.E.; Yu, D.-G.; Ahmad, B.S.; Chatterton, N.P.; Williams, G.R. 5-Fluorouracil loaded Eudragit fibers prepared by electrospinning. *Int. J. Pharm.* **2015**, *495*, 895–902. [CrossRef] [PubMed]
73. Sun, B.; Zhang, M.; Shen, J.; He, Z.; Fatehi, P.; Ni, Y. Applications of cellulose-based materials in sustained drug delivery systems. *Curr. Med. Chem.* **2019**, *26*, 2485–2501. [CrossRef] [PubMed]
74. Roman, M. Toxicity of cellulose nanocrystals: A review. *Ind. Biotechnol.* **2015**, *11*, 25–33. [CrossRef]
75. Xu, H.; Aguilar, Z.P.; Yang, L.; Kuang, M.; Duan, H.; Xiong, Y.; Wei, H.; Wang, A. Antibody conjugated magnetic iron oxide nanoparticles for cancer cell separation in fresh whole blood. *Biomaterials* **2011**, *32*, 9758–9765. [CrossRef] [PubMed]

 *nanomaterials*

Article

Cellulose Nanocrystals/Chitosan-Based Nanosystems: Synthesis, Characterization, and Cellular Uptake on Breast Cancer Cells

Ricardo J. B. Pinto [1], Nicole S. Lameirinhas [1], Gabriela Guedes [1], Gustavo H. Rodrigues da Silva [1], Párástu Oskoei [2], Stefan Spirk [3], Helena Oliveira [2], Iola F. Duarte [1], Carla Vilela [1,*] and Carmen S. R. Freire [1,*]

[1] Department of Chemistry, CICECO—Aveiro Institute of Materials, University of Aveiro, 3810-193 Aveiro, Portugal; r.pinto@ua.pt (R.J.B.P.); nicoleslameirinhas@ua.pt (N.S.L.); gabriela.guedes@ua.pt (G.G.); gustavohrs@ua.pt (G.H.R.d.S.); ioladuarte@ua.pt (I.F.D.)
[2] Department of Biology, CESAM—Centre for Environmental and Marine Studies, University of Aveiro, 3810-193 Aveiro, Portugal; parastu.oskoei@ua.pt (P.O.); holiveira@ua.pt (H.O.)
[3] Institute of Bioproducts and Paper Technology, Graz University of Technology, Inffeldgasse 23, 8010 Graz, Austria; stefan.spirk@tugraz.at
* Correspondence: cvilela@ua.pt (C.V.); cfreire@ua.pt (C.S.R.F.)

Abstract: Cellulose nanocrystals (CNCs) are elongated biobased nanostructures with unique characteristics that can be explored as nanosystems in cancer treatment. Herein, the synthesis, characterization, and cellular uptake on folate receptor (FR)-positive breast cancer cells of nanosystems based on CNCs and a chitosan (CS) derivative are investigated. The physical adsorption of the CS derivative, containing a targeting ligand (folic acid, FA) and an imaging agent (fluorescein isothiocyanate, FITC), on the surface of the CNCs was studied as an eco-friendly methodology to functionalize CNCs. The fluorescent CNCs/FA-CS-FITC nanosystems with a rod-like morphology showed good stability in simulated physiological and non-physiological conditions and non-cytotoxicity towards MDA-MB-231 breast cancer cells. These functionalized CNCs presented a concentration-dependent cellular internalization with a 5-fold increase in the fluorescence intensity for the nanosystem with the higher FA content. Furthermore, the exometabolic profile of the MDA-MB-231 cells exposed to the CNCs/FA-CS-FITC nanosystems disclosed a moderate impact on the cells' metabolic activity, limited to decreased choline uptake and increased acetate release, which implies an anti-proliferative effect. The overall results demonstrate that the CNCs/FA-CS-FITC nanosystems, prepared by an eco-friendly approach, have a high affinity towards FR-positive cancer cells and thus might be applied as nanocarriers with imaging properties for active targeted therapy.

Keywords: cellulose nanocrystals; chitosan; folic acid; fluorescein isothiocyanate; nanosystems; physical adsorption; cellular uptake; cellular exometabolomics; folate receptor-positive cancer cells

check for **updates**

Citation: Pinto, R.J.B.; Lameirinhas, N.S.; Guedes, G.; Rodrigues da Silva, G.H.; Oskoei, P.; Spirk, S.; Oliveira, H.; Duarte, I.F.; Vilela, C.; Freire, C.S.R. Cellulose Nanocrystals/Chitosan-Based Nanosystems: Synthesis, Characterization, and Cellular Uptake on Breast Cancer Cells. *Nanomaterials* **2021**, *11*, 2057. https://doi.org/10.3390/nano 11082057

Academic Editor: Hirotaka Koga

Received: 18 June 2021
Accepted: 9 August 2021
Published: 12 August 2021

Publisher's Note: MDPI stays neutral with regard to jurisdictional claims in published maps and institutional affiliations.

1. Introduction

Polysaccharides are natural polymers composed of monosaccharides linked by glycosidic bonds that hold great promise as eco-friendly building-blocks to develop advanced functional materials [1]. Within the vast portfolio of polysaccharides, cellulose (i.e., linear homopolysaccharide composed of β-D-glucopyranose units linked by β-(1,4) glycosidic bonds) is amongst the most studied natural polymer in multiple fields of research [2–5]. This polysaccharide, and particularly its nanoscale forms, viz. cellulose nanocrystals (CNCs), cellulose nanofibrils (CNFs) and bacterial nanocellulose (BNC) [6], are showing incredible potential as precursors of nanomaterials for a deluge of applications, including in the fight against cancer [7–9]. In fact, the number of developed systems for cancer diagnosis and treatment [7,9,10] is increasing at a fast pace since this large group of diseases is the second leading cause of death globally, with ca. 10 million deaths in 2020 [11,12].

CNCs have a rod-like morphology [13] and are of particular relevance in this context. Recent studies have demonstrated that the shape of the nanocarriers is an important parameter for their efficiency, and that elongated or filamentous nanostructures present several benefits over spherical ones in terms of surface area-to-volume ratio, rate of clearance from the body and elimination mechanism, as well as an enhanced uptake rate by tumor cells [14–17]. Moreover, CNCs are rich in hydroxyl groups that enable easy coupling (covalent and non-covalent) of targeting, imaging, and therapeutic agents [9,18] to actively target cancerous cells [19].

As an example, Roman and co-workers have looked extensively into the use of CNCs as nanocarriers by the functionalization of these elongated nanostructures solely with fluorescein isothiocyanate (FITC) as an imaging agent [20–22] or with both FITC and folic acid (FA) as imaging and targeting agents [23–25], respectively. These authors showed the potential of these nanosystems for the active targeted delivery of chemotherapeutic agents to folate receptor (FR)-positive cancer cells, such as human (DBTRG-05MG, H4) and rat (C6) brain tumor cells [23], KB and human breast cancer cells (MDA-MB-468) [24,25]. Following the same idea, Raja et al. [26] chemically modified CNCs via covalent tethering of PEGylated biotin (targeting ligand) and perylenediimide (imaging agent) and demonstrated their aptness for cell labelling and imaging of fundamental cells of the immune system (J774A.1 macrophages and primary DCs), the connective tissue (NIH-3T3 fibroblasts), and a severe pathological state (HeLa adenocarcinoma cells).

Despite the promising results of the previously enumerated studies, the functionalization of the CNCs nanosystems was mostly performed via multiple synthesis steps and with fairly toxic reagents [20–26], and thus the need for alternative methodologies [18,27] is of utmost importance. Therefore, the use of CNCs in conjunction with a chitosan (CS, i.e., linear heteropolysaccharide obtained from chitin via *N*-deacetylation [28]) derivative with targeting and imaging ligands seems an interesting approach (inspired by their opposite surface charge) that has never been addressed before. In this scenario, the present study describes the synthesis, characterization, and cellular uptake on breast cancer cells (MDA-MB-231 cell line) of a dual polysaccharide nanosystem based on CNCs and a CS derivative with opposite surface charges. The physical adsorption of the chitosan derivative, containing targeting (FA) and imaging (FITC) ligands, on the surface of the CNCs was studied as an environmentally friendly approach to engineer CNCs nanosystems with enhanced cellular internalization by MDA-MB-231 cells. The CNCs nanosystems were characterized in terms of structure, optical properties, fluorescence, stability in simulated physiological and non-physiological conditions, morphology, in vitro cytotoxicity, cellular internalization, and extracellular metabolomic profile to evaluate their suitability for the targeting and imaging of breast cancer cells.

2. Materials and Methods

2.1. Chemicals, Materials, and Cells

Acetic acid (97.0%, Sigma-Aldrich (St. Louis, MO, USA)), chitosan (CS, purified powder, MW 15,000, >85% degree of deacetylation, viscosity >90 cPs at 1% solution, Polysciences (Warrington, PA, USA)), Azpack™ cotton wool (BP grade, Fisher Scientific (Hampton, NH, USA)), dimethyl sulfoxide (DMSO, >99%, LabScan (Bangkok, Thailand)), 3-(4,5-dimethylthiazolyl-2)-2,5-diphenyltetrazolium bromide (MTT, 98%, Sigma-Aldrich (St. Louis, MO, USA)), 1-ethyl-3-(3-dimethylaminopropyl)carbodiimide (EDC, ≥98.0%, Sigma-Aldrich (St. Louis, MO, USA)), fluorescein isothiocyanate (FITC, isomer I, 90%, Sigma-Aldrich (St. Louis, MO, USA)), folic acid (FA, 97.0%, Sigma-Aldrich (St. Louis, MO, USA)), methanol (99.9%, Fisher Chemical (Hampton, NH, USA)), phosphate buffered saline (PBS, pH 7.2, Sigma-Aldrich (St. Louis, MO, USA)), sodium hydroxide (99.0%, Fisher Chemical (Hampton, NH, USA)), and phosphotungstic acid hydrate (HPW, p.a., Carl Roth (Karlsruhe, Germany)) were used as received. Other chemicals and solvents were of laboratory grade.

RPMI 1640 medium, fetal bovine serum (FBS), L-glutamine, penicillin/streptomycin, fungizone (250 U mL^{-1}) and trypsin–ethylenediaminetetraacetic acid (EDTA) (0.25% trypsin and 1 mM EDTA) were purchased from Gibco® (Life Technologies, Carlsbad, CA, USA). Type 1 ultrapure water (resistivity of 18.2 MΩ cm (25 °C)) was filtered by a Simplicity® Water Purification System (Merck Millipore, Darmstadt, Germany). The Spectra/Por® regenerated cellulose dialysis membranes (MWCO 6–8 kDa) were purchased from Daigger Scientific (Hamilton, NJ, USA). The MDA-MB-231 breast cancer cells were obtained from the American Type Culture Collection (ATCC, Manassas, VA, USA).

2.2. Preparation of Cellulose Nanocrystals (CNCs)

The CNCs were prepared via acid hydrolysis with phosphotungstic acid according to a modified literature protocol [29]. In a typical procedure, 322 g of HPW (111 mmol) and water (110 mL) were placed in a 500 mL three-neck round bottom flask equipped with a reflux condenser. Then, the mixture was heated to 90 °C and thoroughly washed cellulose cotton fibers (9.0 g, repeated washing with distilled water to remove impurities) were added in small portions. To achieve hydrolysis of the cotton fibers, the mixture was stirred for 48 h under reflux. After cooling to room temperature under vigorous stirring, the mixture was transferred to a 3 L extraction funnel. Diethyl ether (500 mL) and water (250 mL) were added, resulting in the formation of three phases (heaviest phase: HPW/ethanol/water, medium phase: aqueous, upper phase: ethanol). The medium aqueous phase was separated, and the CNCs were purified by repeated centrifugation and washing with NaOH (0.1 M) and distilled water to remove traces of tungstate. The average yield after freeze-drying (5 experiments) was 86.3 ± 3.1%.

The obtained CNCs have an elemental composition of 40.3 ± 0.4% of carbon (C), 5.4 ± 0.1% of hydrogen (H) and 54.3 ± 0.5% of oxygen (O) determined by elemental analysis (LECO TruSpec 630-200-200 CHNS, LECO Corporation, Michigan, USA), dimensions of 45 ± 12 nm (diameter) and 364 ± 74 nm (length) (micrographs acquired on a HR-SEM-SE SU-70 microscope, Hitachi High-Technologies Corporation, Tokyo, Japan), and zeta (ζ)-potential of −11.5 ± 0.7 mV at pH 7 (Zetasizer Nano ZS, Malvern Panalytical, Cambridge, UK).

2.3. Synthesis of the FA-CS-FITC Derivative

Firstly, CS (ζ-potential of +68.4 ± 2.1 mV at pH 3) was functionalized with folic acid (i.e., CS-FA derivative) following the methodology described by Lee et al. [30] with some modifications. Briefly, EDC (20 mg) was added to an FA solution (20 mg in 10 mL DMSO) and allowed to react for 90 min at room temperature to activate the carboxylic groups of FA. This solution was then added dropwise to a CS solution (100 mg in 20 mL of 1 M acetic acid aqueous solution) under magnetic stirring, in an ice bath, and the pH was adjusted to 6 using an aqueous solution of NaOH (1 M). The reaction mixture was kept under stirring over 48 h at room temperature in the dark. The CS-FA derivative was dialyzed in distilled water for three days to remove the unreacted FA, followed by freeze-drying, which generated a yellowish solid with an elemental composition of 47.9 ± 0.4% of C, 6.2 ± 0.3% of H, 33.0 ± 0.5% of O, and 12.9 ± 0.3% of nitrogen (N), and a ζ-potential value of +57.0 ± 4.4 mV at pH 3.

The FA-CS-FITC derivative was synthesized via an adapted method described by Huang et al. [31]. A solution of FITC (12 mg in 18 mL of methanol) was added dropwise to a solution of CS-FA (60 mg in 18 mL of 1 M acetic acid aqueous solution) under magnetic stirring. The reaction mixture was allowed to react for 24 h at room temperature in the dark. Then, the pH of the solution was adjusted to 10 with an aqueous solution of NaOH (1 M). The solution containing the FA-CS-FITC derivative was dialyzed in distilled water for four days, followed by freeze-drying, which produced a light orange solid with an elemental composition of 50.0 ± 0.5% of C, 5.3 ± 0.3% of H, 37.5 ± 0.4% of O, 6.6 ± 0.1% of N and 0.6 ± 0.1% of sulphur (S), and a ζ-potential value of +51.9 ± 3.6 mV at pH 3.

2.4. Preparation of the CNC Nanosystems Functionalized with the FA-CS-FITC Derivative

The CNCs were functionalized with the FA-CS-FITC derivative via electrostatic assembly. Briefly, a solution of FA-CS-FITC derivative (1, 2 or 4 mg in 40 mL of 1 M acetic acid aqueous solution) was added dropwise to a suspension of CNCs (50 mg in 10 mL of ultrapure water), as summarized in Table 1. The mixture stood for 1 h under magnetic stirring in the dark. Afterwards, the suspension was centrifuged (13,400 rpm, 10 min, Megafuge 16R centrifuge (Thermo Scientific, Waltham, MA, USA)) and washed one time with 1 M acetic acid aqueous solution and three times with ultrapure water, followed by freeze-drying, which yielded solid nanosystems with an orange coloration.

Table 1. Composition of the prepared CNCs/FA-CS-FITC nanosystems.

Nanosystems	Nominal Composition [a]		Measured Composition [b]	
	W_{CNCs} (mg)	$W_{FA\text{-}CS\text{-}FITC}$ (mg)	$W_{FA\text{-}CS\text{-}FITC}$ (mg)	$W_{FA\text{-}CS\text{-}FITC}/W_{CNCs}$
CNCs/FA-CS-FITC_1	50.0	1.01	0.65	0.013
CNCs/FA-CS-FITC_2	50.0	1.99	1.43	0.029
CNCs/FA-CS-FITC_3	50.0	4.09	2.97	0.059

[a] The nominal composition is the initial mass of CNCs (W_{CNCs}) and CS derivative ($W_{FA\text{-}CS\text{-}FITC}$). [b] The measured composition was indirectly determined by measuring the absorbance of the washing solutions of each nanosystem at 441 nm (calibration curve: $y = 2.2591x - 0.0141$, $R^2 = 0.9995$), concentration range of the FA-CS-FITC derivative: 0.005–0.5 mg mL^{-1}).

2.5. Characterization Methods

The Fourier transform infrared-attenuated total reflection (FTIR-ATR) spectra of all samples were collected in the solid-state in a Perkin-Elmer FT-IR System Spectrum BX spectrophotometer (Perkin-Elmer, Waltham, MA, USA), coupled with a single horizontal Golden Gate ATR cell (Specac®, London, UK), using 32 scans at a resolution of 4 cm^{-1} in the wavenumber range of 600–4000 cm^{-1}.

The optical spectra were recorded on a Thermo Scientific Evolution 220 UV-visible spectrophotometer (Thermo Fisher Scientific, Waltham, MA, USA) using 100 scans min^{-1} with a bandwidth of 2 nm and an integration time of 0.3 s in the wavelength range of 250–600 nm. The CS and the FA-CS-FITC samples were analyzed in acidic aqueous solutions (1 M of acetic acid), and the FA and FITC samples were examined in DMSO and ethanol (1 mg mL^{-1}), respectively. For the solid samples, namely the CNCs and CNCs/FA-CS-FITC_3 nanosystem, a Jasco V-560 UV-visible spectrophotometer (JASCO Corporation, Tokyo, Japan) was utilized also in the wavelength range of 250–600 nm.

The fluorescence emission spectra were obtained on a Horiba Jobin-Yvon FluoroMax-4 spectrofluorometer (Horiba Jobin-Yvon, Kyoto, Japan) with a 2.0 nm width of both excitation and emission slits, and an integration time of 0.1 s in the wavelength range of 500–650 nm. The appropriate excitation wavelengths (λ_{ex}) were obtained as the wavelength of maximum absorption as follows: λ_{ex} = 450 nm for the FITC sample in ethanol (1 mg mL^{-1}) and λ_{ex} = 440 nm for the FA-CS-FITC in an acidic aqueous solution (ca. 1 mg mL^{-1}, 1 M of acetic acid). The fluorescence spectrum of the solid CNCs/FA-CS-FITC_3 nanosystem was recorded using a Jasco spectrofluorometer FP-8300 (JASCO Corporation, Tokyo, Japan) equipped with a xenon lamp, using a scan speed of 200 nm min^{-1}, bandwidth of excitation and emission of 5 nm and excitation wavelength of 495 nm, also in the wavelength range of 500–650 nm.

Zeta potential measurements were performed on a Malvern ZetaSizer Nano-ZS (Malvern Panalytical, Cambridge, UK) at room temperature and different media depending on the analyzed sample, as described above. All measurements were performed in triplicate.

Scanning transmission electron microscopy (STEM) images were acquired in a field-emission gun (FEG) SEM Hitachi SU-70 microscope (Hitachi High-Technologies Corporation, Tokyo, Japan) operated at 15 kV. Samples were prepared by placing a drop of the suspensions of CNCs and CNCs/FA-CS-FITC_3 nanosystem directly onto carbon-coated copper grids and allowing the solvent to evaporate. The size (diameter and length) of the samples was determined by measuring over 100 rods (elongated nanostructures) for each sample, using the Fiji image processing software.

2.6. Stability Tests

The stability of the CNCs/FA-CS-FITC nanosystems was evaluated in simulated non-physiological and physiological conditions, namely at pH 2.1 (aqueous solution of 0.01 M HCl) and at pH 7.2 (PBS), for 24 h and 48 h [32]. Typically, 1 mg of CNCs/FA-CS-FITC was added to vials containing 2 mL of the two media. Then, the suspensions were placed on an orbital shaker at 37 °C in the dark for 24 h and 48 h. After these periods, the suspensions were centrifugated at 13,400 rpm during 5 min (Megafuge 16R centrifuge, Thermo Scientific, Waltham, MA, USA). The absorbance spectra of the corresponding supernatants were recorded on a Thermo Scientific Evolution 220 UV-visible spectrophotometer (Thermo Scientific, Waltham, MA, USA) to quantify the amount of the derivative that was released (calibration curve: $y = 2.2591x - 0.0141$, $R^2 = 0.9995$), concentration range of the FA-CS-FITC derivative: 0.005–0.5 mg mL^{-1}).

2.7. Cell Culture

MDA-MB-231 cells were cultured in RPMI culture medium with L-glutamine, without folic acid, supplemented with 10% FBS, 2 mM L-glutamine, 1% penicillin-streptomycin (10,000 U mL^{-1}), and 1% fungizone (250 U mL^{-1}), at 37 °C in a humidified atmosphere with 5% CO_2. Cells were daily observed for confluence and morphology using an inverted phase-contrast Eclipse TS100 microscope (Nikon, Tokyo, Japan). Sub-confluent cells were trypsinized with trypsin-EDTA (0.25% trypsin, 1 mM EDTA) when monolayers reached 70% of confluence.

2.8. In Vitro Cytotoxicity Assay

Cells were seeded in a 96-well plate at 20,000 cells/well and, after cell adhesion, the cell culture medium (in the 96 well plates) was replaced with fresh medium containing CNCs and CNCs/FA-CS-FITC nanosystems at 0, 12.5, 25, 50, 100, 200 µg mL^{-1} and then further incubated for 24 h at 37 °C, 5% CO_2 humidified atmosphere.

At the end of the exposure time, 50 µL of MTT solution (1 mg mL^{-1} in PBS pH 7.2) were added to the medium and cells were incubated for 4 h. After that, the culture medium with MTT was removed and replaced by 150 µL of DMSO and the plate was placed in a shaker for 2 h in the dark to completely dissolve the formazan crystals. The absorbance of the samples was measured with a BioTek Synergy HT plate reader (Synergy HT Multi-Mode, BioTeK, Winooski, VT, USA) at 570 nm with blank corrections. The cell viability was calculated with respect to the control cells:

$$Cell\ viability\ (\%) = \left[\left(Abs_{sample} - Abs_{DMSO} \right) / \left(Abs_{control} - Abs_{DMSO} \right) \right] \times 100 \quad (1)$$

where Abs_{sample} is the absorbance of the sample, Abs_{DMSO} is the absorbance of the DMSO solvent and $Abs_{control}$ is the absorbance of the control.

2.9. Cellular Uptake Assay

Cells were seeded in 24-well plates at the concentration of 138,320 cells/well and then incubated for 24 h at 37 °C and 5% CO_2, for cell adherence. The culture medium was then replaced with 500 µL of growth medium with pristine CNCs and CNCs/FA-CS-FITC nanosystems at 200 µg mL^{-1}, and the cells were incubated with the previously described culture conditions. Acellular medium and control cells (no treatment) were also incubated. After the 24 h incubation period, the medium was collected and 400 µL aliquots were stored at −80 °C. Following washing with 250 µL of PBS, the cells were trypsinized and 250 µL of growth medium was added to neutralize trypsin. Two parameters, namely side-scattered light, and side-fluorescence light (excitation at 488 nm and measurement with a 530/30 band pass filter), were measured in an Attune® Acoustic Focusing Cytometer

(ThermoFisher Scientific, Waltham, MA, USA). At least 50,000 cells were examined for each test. The data were analyzed by FlowJo software (FlowJo LLC, Ashland, OR, USA).

2.10. Cellular Exometabolomics

For the NMR analysis, the medium samples collected during the cellular uptake assay (described in Section 2.9) were processed to remove interfering proteins. Briefly, 700 µL of cold methanol were added to 350 µL of medium, followed by 30 min resting at $-20\ ^{\circ}$C, centrifugation (13,000$\times$ g, 20 min) and vacuum drying of the supernatant. Dried samples were then reconstituted in 600 µL of deuterated PBS (100 mM, pH 7.4) and transferred into 5 mm NMR tubes.

The NMR spectra were acquired on a Bruker Avance III HD 500 NMR spectrometer (Bruker Corporation, Billerica, MA, USA) operating at 500.13 MHz for ^{1}H observation using a 5 mm TXI probe. Standard 1D ^{1}H spectra with water pre-saturation (pulse program 'noesypr1d', Bruker library) were recorded with 32 k points, 7002.801 Hz spectral width, a 2 s relaxation delay and 512 scans. Spectral processing (TopSpin 4.0.3, Bruker BioSpin) comprised cosine multiplication (ssb 2), zero-filling to 64 k data points, manual phasing, baseline correction, and calibration to the TSP-$_{d4}$ signal (δ 0 ppm). Selected signals representative of the main metabolites detected were then integrated (Amix-Viewer 3.9.15, Bruker Biospin) and, for each metabolite, the fold change relative to the acellular medium was calculated for control and exposed groups to assess the magnitude of consumptions/secretions.

2.11. Statistical Analysis

Cellular viability, uptake and exometabolomics data were analyzed using the Graph-Pad Prism Software (GraphPad Software Inc., San Diego, CA, USA) with the data presented as the mean values $\pm$ standard error. Where differences existed, the source of the differences at the $p < 0.05$ significance level was identified by all pairwise multiple comparison procedures via the Tukey's test.

3. Results and Discussion

Nanosystems based on cellulose nanocrystals (CNCs) and a multifunctional chitosan (CS) derivative were prepared via the simple (and eco-friendly) physical adsorption of the CS derivative, containing targeting (folic acid, FA) and imaging (fluorescein isothiocyanate, FITC) ligands (i.e., FA-CS-FITC), on the surface of the CNCs (Figure 1). Herein, the FA vitamin was carefully chosen as a targeting agent given the over-expression of folate receptors (FRs) in several tumor cells (and under-expression in non-tumor cells) [9,33], while the FITC fluorophore was picked for being a widely used imaging probe for the flow cytometry assays [33]. Furthermore, the CNCs were selected for their elongated nanostructure and anionic surface charge [29], whereas the CS derivative was chosen for its imaging and targeting moieties, as well as the cationic surface charge due to the presence of the characteristic protonated amine groups of CS at an acidic pH [34]. The assembly of the CNCs/FA-CS-FITC nanosystems via non-covalent interactions was preferred to covalent bonding due to the milder reaction conditions, when compared with previous studies where the functionalization of the CNC nanosystems was performed with fairly toxic reagents [20,22,23,25].

The opposite surface charge of these two polysaccharides enabled the facile physical adsorption [35] of the multifunctional CS derivative (i.e., FA-CS-FITC) on the surface of the elongated CNCs (Figure 1). The CNCs/FA-CS-FITC nanosystems were characterized in terms of structure, optical properties, fluorescence, stability in simulated physiological and non-physiological conditions, morphology, in vitro cytotoxicity, cellular internalization, and exometabolomics profile to evaluate their suitability for the targeting and imaging of breast cancer cells.

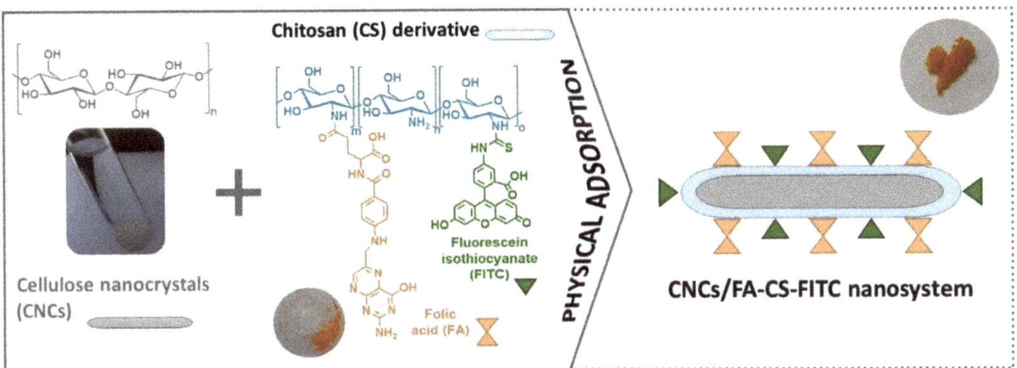

Figure 1. Scheme representing the preparation of the CNCs/FA-CS-FITC nanosystems.

3.1. Preparation and Characterization of the CNCs/FA-CS-FITC Nanosystems

The synthetic pathway to obtain the CNCs nanosystems with targeting and imaging functions (Figure 1) comprised three main steps. First, the CNCs were extracted from cellulose cotton fibers via acid hydrolysis with phosphotungstic acid [29]. Then, the FA-CS-FITC derivative was synthesized via a two-step pathway, where the CS was initially functionalized with FA by a carbodiimide-mediated amidation reaction, followed by the reaction of this intermediate with the isothiocyanate groups (N=C=S) of the FITC to obtain the FA-CS-FITC derivative [30,31]. Lastly, the simple and eco-friendly physical adsorption of the FA-CS-FITC derivative (ζ-potential of +51.9 ± 3.6 mV at pH 3) on the surface of the CNCs (ζ-potential of −11.5 ± 0.7 mV at pH 7) generated the CNCs/FA-CS-FITC nanosystems (Figure 1). Herein, three CNCs nanosystems with different contents of the FA-CS-FITC derivative were prepared, namely CNCs/FA-CS-FITC_1 with a content of 13 μg of FA-CS-FITC per mg of CNCs, CNCs/FA-CS-FITC_2 with 29 μg mg^{-1} and CNCs/FA-CS-FITC_3 with 59 μg mg^{-1} (Table 1). The adsorption efficiencies of the FA-CS-FITC derivative on the surface of the CNCs (determined by UV-vis spectroscopy) were ca. 64% for CNCs/FA-CS-FITC_1, and around 72% for both CNCs/FA-CS-FITC_2 and CNCs/FA-CS-FITC_3 (Table 1).

The first indication of the successful assembly process between the CNCs and the CS derivative was given by the color change of the elongated CNCs from white to an orange color, as illustrated in Figure 1. This was further corroborated by FTIR-ATR, UV-vis, and fluorescence spectroscopy, as exemplified in Figure 2 for the nanosystem with the higher content of the FA-CS-FITC derivative, i.e., CNCs/FA-CS-FITC_3 (Table 1).

The FTIR-ATR spectrum of the CNCs/FA-CS-FITC_3 nanosystem (Figure 2B) presents predominantly the characteristic vibrations of the CNCs at 3342 cm^{-1} (O–H stretching), 2902 cm^{-1} (C–H stretching), 1316 cm^{-1} (O–H in-plane bending), and 1032 cm^{-1} (C–O stretching) (Figure 2B) [36], but also those of the CS derivative, namely from: (i) chitosan at 3342 cm^{-1} (O–H and N–H stretching), 1636 cm^{-1} (C=O stretching and N–H bending), 1592 cm^{-1} (–NH$_2$ bending), 1380 cm^{-1} (–CH$_2$ bending), 1078 and 1032 cm^{-1} (C–O stretching) [37]; (ii) folic acid at 1688 cm^{-1} (C=O stretching), 1602 and 1481 cm^{-1} (C=C aromatic) cm^{-1} (phenyl ring) [38], and (iii) FITC at 1545, 1458 and 1378 cm^{-1} (xanthene ring skeletal C–C stretching) [39–41], as depicted in Figure 2A. However, the majority of these absorption bands overlap because of the common functional groups of the individual components (i.e., CNCs, CS, FA and FITC). Worth noting here is the absence of the absorption bands at 1686 cm^{-1} (carboxyl moiety of FA [38]) and 2018 cm^{-1} (isothiocyanate moiety (N=C=S) of the FITC [39]), together with the appearance of the absorption band (ca. 1650 cm^{-1}) assigned to the amide thioamide bonds [42], that confirm the covalent link between CS and FA [43], and CS and FITC [44], respectively, in agreement with data reported elsewhere [45].

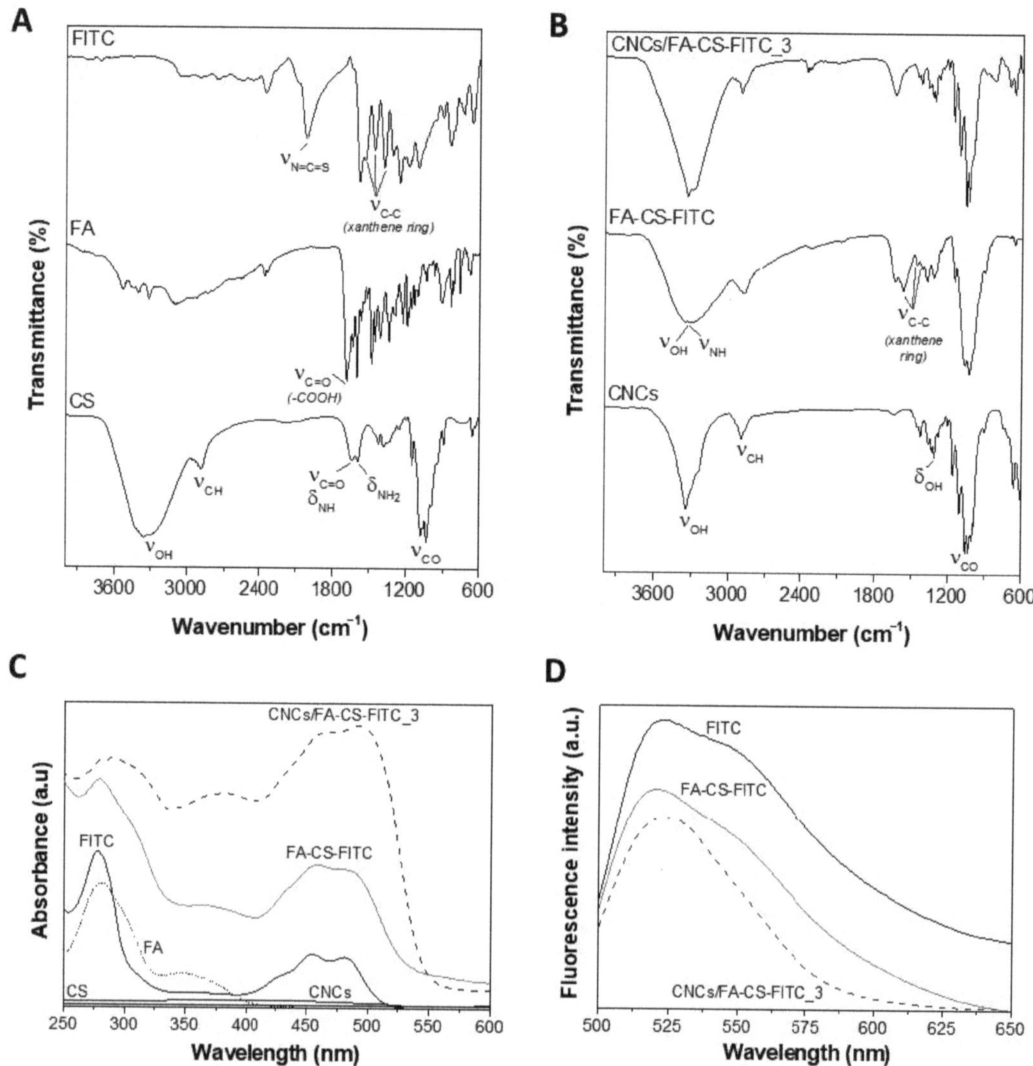

Figure 2. (**A**) FTIR-ATR spectra (vibrational modes: ν = stretching, δ = bending) of the pure CS, FA and FITC, (**B**) FTIR-ATR spectra of the pristine CNCs, FA-CS-FITC derivative and CNCs/FA-CS-FITC_3 nanosystem, (**C**) UV-vis spectra of CNCs, CS, FA, FITC, FA-CS-FITC derivative and CNCs/FA-CS-FITC_3 nanosystem, and (**D**) fluorescence emission spectra of the FITC, FA-CS-FITC derivative and CNCs/FA-CS-FITC_3 nanosystem.

The UV-vis spectrum of the CNCs/FA-CS-FITC_3 nanosystem (Figure 2C) exhibits only the characteristic bands of the CS derivative, given that the CNCs do not present any absorption in the UV and visible regions, as registered elsewhere [20,46]. On the other hand, the bands of the FA-CS-FITC derivative (Figure 2C) are typical of the FA (at ca. 285 and 356 nm ascribed to the π–π^* and n–π^* transitions, respectively [38]) and FITC (at about 289, 462 and 490 nm [39,40]) ligands since the pure CS polysaccharide does not display any absorption in the UV and visible regions [47]. Predictably, some of the absorption bands were red-shifted compared to those reported for the pure FA and FITC, which is an indication of the covalent bond between CS, FA and FITC in the FA-CS-FITC derivative, as described for other systems such as FA-polyaniline [38] and FITC-organoclay [39].

The fluorescence emission spectrum of the CNCs/FA-CS-FITC_3 nanosystem (Figure 2D) shows an absorption maximum at ca. 524 nm, which mimics the emission fluorescence peak of the FA-CS-FITC derivative. This peak is characteristic of the FITC fluorophore that, as reported by Ghosh et al. [40], upon excitation at a wavelength of 490 nm presents a strong emission peaking at around 520 nm. As anticipated, the pristine CNCs showed no emission in the wavelength range of the visible region. These results indicate that the CNCs nanosystems have fluorescent properties since the FITC fluorophore maintained its fluorescence. This behavior was expected given that the FITC is widely used to attach a fluorescent label mostly to polysaccharides (e.g., FITC-labelled CS [33,48], FITC-labelled CNCs [20,49], FITC-labelled dextran [40]) and proteins (e.g., FITC-labelled albumin [50]).

The success of the assembly process was additionally confirmed by testing the stability of the CNCs/FA-CS-FITC_3 nanosystem under simulated non-physiological and physiological pH conditions, namely pH 2.1 for acidic medium [32] and pH 7.2 for human blood plasma [51], respectively. Since these CNC nanosystems rely on the physical adsorption between the negatively charged CNCs and the cationic CS derivative, it is important to assess the potential release of FA-CS-FITC from the nanosystems to avoid ambiguous interpretations of the results or even unnecessary cytotoxic effects, which might hinder their applicability [52]. According to the obtained data, the amount of the CS derivative released from the CNCs/FA-CS-FITC_3 nanosystem is low at a non-physiologic pH with values of 6.6 ± 0.8% and 8.4 ± 0.5% for 24 h and 48 h, respectively. At the physiological pH 7.2, the release was slightly higher, with values of 11.0 ± 0.1% and 14.5 ± 0.1% for 24 h and 48 h, respectively. Overall, the CNCs/FA-CS-FITC nanosystems present good stability and dispersibility in both pH conditions, which can be credited to the effective electrostatic interactions established between the anionic CNCs and the cationic FA-CS-FITC derivative.

The morphology and size of the CNCs nanosystems functionalized with the FA-CS-FITC derivative were assessed by STEM (Figure 3). As anticipated, the pristine CNCs showed the typical rod-like morphology with dimensions of 45 ± 12 nm (diameter) and 364 ± 74 nm (length), analogous to the data described in the literature [29]. After the assembly process, the CNCs maintained the rod-like morphology, but their dimensions slightly increased to 57 ± 15 nm (diameter) and 442 ± 124 nm (length). This minor increase in the size range is clearly a direct result of the physical adsorption of the CS derivative on the surface of the CNCs, and it was also observed for other CNCs-based nanosystems [25,53].

Both the shape and size of the CNCs/FA-CS-FITC nanosystems are parameters of utmost importance due to the cell-size-dependent uptake of nanoparticles by the cells [14,17]. Thus, one can speculate that the elongated morphology of the CNCs/FA-CS-FITC nanosystems, together with their nanometric size, will not hinder the cellular internalization by the FR-positive breast cancer cells (i.e., MDA-MB-231 cell line), as will be discussed in the following paragraphs.

3.2. Cellular Viability

The in vitro cytotoxicity of the pristine CNCs and the CNCs/FA-CS-FITC nanosystems was evaluated in human breast adenocarcinoma cells (i.e., MDA-MB-231 cells) for 24 h at concentrations ranging from 12.5 to 200 μg mL^{-1} using the indirect MTT assay. The MDA-MB-231 cell line was selected for being often used as a model of triple-negative breast cancer cells [54] with over-expression of folate receptors [55], and for which the available therapeutic options are more scarce than other breast cancer subtypes [56].

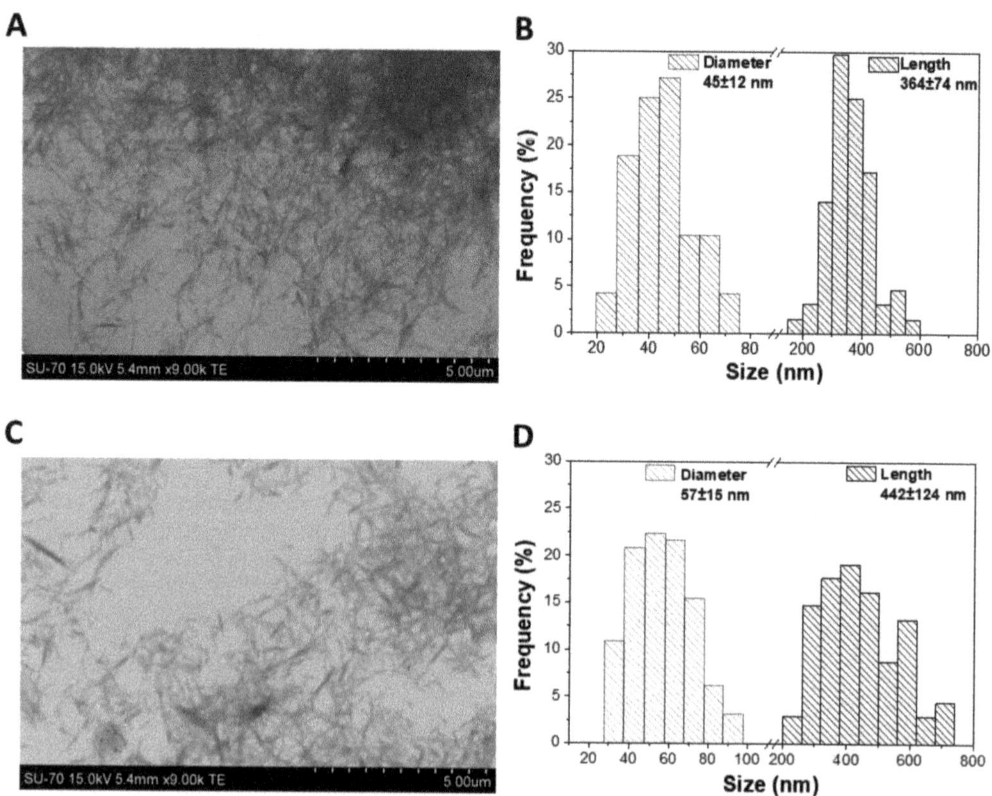

Figure 3. STEM micrographs (**A,C**) with the corresponding size histograms (diameter and length, (**B,D**)) of the (**A,B**) pristine CNCs and (**C,D**) CNCs/FA-CS-FITC_3 nanosystem.

According to the data presented in Figure 4A, the cell viability of the MDA-MB-231 cells after 24 h of exposure to the pristine CNCs at five different concentrations (12.5, 25, 50, 100 and 200 µg mL^{-1}) remained at the level of 100%, meaning that the cell viability is not dose-dependent in the tested concentration range. This outcome for the pristine CNCs is consistent with the results found in the literature for this cell line [22], but also for other FR-positive human cancer cell lines, e.g., MDA-MB-468, KB, and PC-3 cells [22].

Regarding the CNCs/FA-CS-FITC nanosystems, the profile is the same and the cell viability is higher than 90% for all the tested concentrations. This confirms without a doubt that the three CNCs/FA-CS-FITC nanosystems are non-cytotoxic to the MDA-MB-231 cells at concentrations below 200 µg mL^{-1}, with an in vitro cell viability quite above the 70% threshold [57]. These results are similar to those reported for other CNCs-based nanosystems with other tumor cell lines, such as the human breast adenocarcinoma MCF-7 cell line [58], and the A375 and M14 cells (malignant melanoma cell lines) [59].

The non-cytotoxicity of the three CNCs/FA-CS-FITC nanosystems for the tested concentration range evinces their potential to act as nanocarriers of imaging and therapeutic agents for targeted therapy, without inhibiting the viability of the FR-positive breast cancer cells. In this sense, the cellular internalization assays will be performed with the highest tested concentration, namely at 200 µg mL^{-1} of each of the CNCs/FA-CS-FITC nanosystems, as discussed in the following paragraphs.

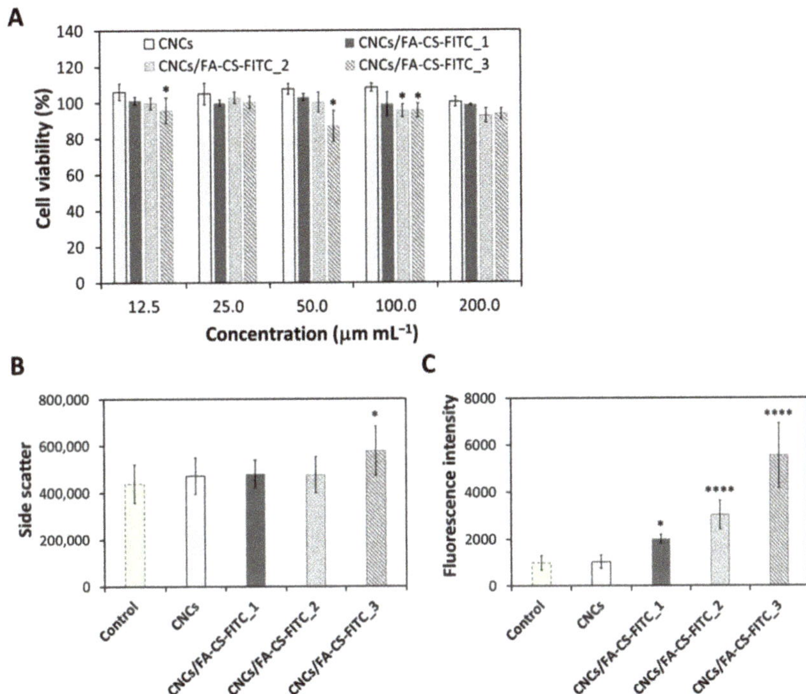

Figure 4. (**A**) Cell viability of the MDA-MB-231 cells after 24 h of exposure to the pristine CNCs and CNCs/FA-CS-FITC nanosystems, and flow cytometry data: (**B**) side-scattered light and (**C**) side-fluorescence light of the pristine CNCs and CNCs/FA-CS-FITC nanosystems for 24 h (the symbols * and **** epitomize the means with a significant difference from the control at $p < 0.05$ and $p < 0.0001$ levels, respectively).

3.3. Cellular Internalization

The in vitro cellular internalization of the CNCs/FA-CS-FITC nanosystems by the MDA-MB-231 cells was studied by flow cytometry (side-scattered light (Figure 4B) and side-fluorescence light (Figure 4C)) for an exposure time of 24 h. The data compiled in Figure 4B shows that the incubation of the MDA-MB-231 cells in the presence of the pristine CNCs did not result in cell internalization, which is in line with the findings reported for this cell line [22], but also for other FR-positive human cancer cell lines, namely KB and MDA-MB-468 cells [22,24,25], and PC-3 cells [22].

In the case of the CNCs/FA-CS-FITC nanosystems, the ones with the lower content of the CS derivative, i.e., CNCs/FA-CS-FITC_1 (13 µg of FA-CS-FITC per mg of CNCs) and CNCs/FA-CS-FITC_2 (29 µg of FA-CS-FITC per mg of CNCs), showed a minor enhancement when compared to the control, and thus the cell internalization was minimal. In fact, only the CNCs/FA-CS-FITC_3 nanosystem (59 µg of FA-CS-FITC per mg of CNCs) was significantly internalized by the FR-positive breast cancer cell line ($p < 0.05$).

Since these measurements are proportional to cell granularity or cell complexity and thus can be sometimes less sensitive [60], the fluorescence intensity of the cell population was also evaluated (Figure 4C) by taking advantage of the existent FITC fluorophore probe in the CNCs/FA-CS-FITC nanosystems. Once again, the cells incubated in the presence of the pristine CNCs exhibited a fluorescence intensity matching the control, which agrees with the results shown in Figure 4B. On the contrary, the CNCs/FA-CS-FITC nanosystems displayed a significant increase in the fluorescence intensity (Figure 4C). Basically, it is possible to observe the augment of the fluorescence intensity with the increasing content of the CS derivative (i.e., FA-CS-FITC, Table 1), reaching a maximum 5-fold increase compared to the control. The growing cell internalization by the FR-positive breast cancer cells is credited to the increasing content of FA.

Notably, this outcome also validates that the size of the CNCs/FA-CS-FITC nanosystems (Figure 3) did not hamper in any way the effective internalization of these FA-target nanosystems by the FR-positive breast cancer cells, as is in fact shown with other CNCs-based nanosystems with FA ligands [23,25,26].

3.4. Cellular Exometabolomics

The extracellular metabolomic profile of the MDA-MB-231 cells was studied by ^{1}H NMR spectroscopy to assess possible changes in the cells metabolic activity induced by the 24 h exposure to the pristine CNCs and the CNCs/FA-CS-FITC nanosystems (200 µg mL^{-1}). Figure 5A shows the characteristic ^{1}H NMR spectra of the MDA-MB-231 cell-conditioned medium and of the acellular culture medium incubated under identical conditions. Integration of the NMR signals representative of the main metabolites allowed the determination of the metabolite consumption and secretion by control and exposed cells, as summarized in Figure 5B.

The MDA-MB-231 control cells mainly consumed glucose and choline, along with some amino acids (glutamine, branched-chain amino acids, aromatic amino acids, and histidine), while excreting lactate, glutamate, alanine, and acetate (Figure 5B,C). These findings are similar to those reported by Guerra et al. [61] and highlight the fact that the MDA-MB-231 cell line, like most cancer cells, exhibits the classical Warburg effect with a high glucose uptake and lactate production, together with high glutaminolytic activity [62].

Upon exposure of the MDA-MB-231 cells to the CNCs, there were no changes in their metabolic activity, which is not unexpected, given that the pristine CNCs were poorly internalized (Figure 4B,C). However, when cells were incubated with the CNCs/FA-CS-FITC nanosystems, slight but significant dose-dependent differences were found in choline uptake and acetate secretion (Figure 5B). The CNCs functionalized with intermediate and high FA concentrations (i.e., CNCs/FA-CS-FITC_2 and CNCs/FA-CS-FITC_3, respectively) caused a decrease in choline consumption and an increase in acetate release to the extracellular medium. Choline is an essential vitamin-like nutrient required for the de novo synthesis of membrane phospholipids, like phosphatidylcholine and sphingomyelin [63]. Various types of cancer cells, including breast cancer, display enhanced choline uptake and altered metabolism, to support fast proliferation and migratory capacity [64]. Hence, it is possible that the herein observed decrease in choline consumption by the cells incubated with the CNCs/FA-CS-FITC nanosystems reflects a lower synthesis of membrane lipids and a slower proliferation. This is consistent with the observed increase in the amount of acetate released by cells upon incubation with the nanosystems containing the highest concentrations of the FA-CS-FITC derivative (i.e., CNCs/FA-CS-FITC_2 and CNCs/FA-CS-FITC_3, Table 1). Indeed, besides glucose and glutamine, acetate provided as an extracellular nutrient or produced endogenously from pyruvate [65], may chiefly contribute to de novo lipid synthesis via conversion into acetyl-CoA by cytosolic and/or mitochondrial acetyl-CoA synthetases [66]. Hence, its release into the culture medium could reflect its lower intracellular utilization, possibly in relation to the downregulation of lipogenesis.

To summarize, the assembly of the CNCs/FA-CS-FITC nanosystems via physical adsorption was inspired by the possibility of using the opposite surface charge of the CNCs and the FA-CS-FITC derivative and originated non-cytotoxic nanosystems up to concentrations of 200 µg mL^{-1}. Furthermore, and since the cellular uptake is a limiting factor for the efficacy of countless anticancer drugs, the presence of FA in the CNCs/FA-CS-FITC nanosystems promoted a higher cellular internalization towards FR-positive MDA-MB-231 breast cancer cells. Besides, mild alterations in the cells' exometabolome upon 24 h exposure to the CNCs/FA-CS-FITC nanosystems suggest an anti-proliferative effect, which may be beneficial in the context of cancer treatment.

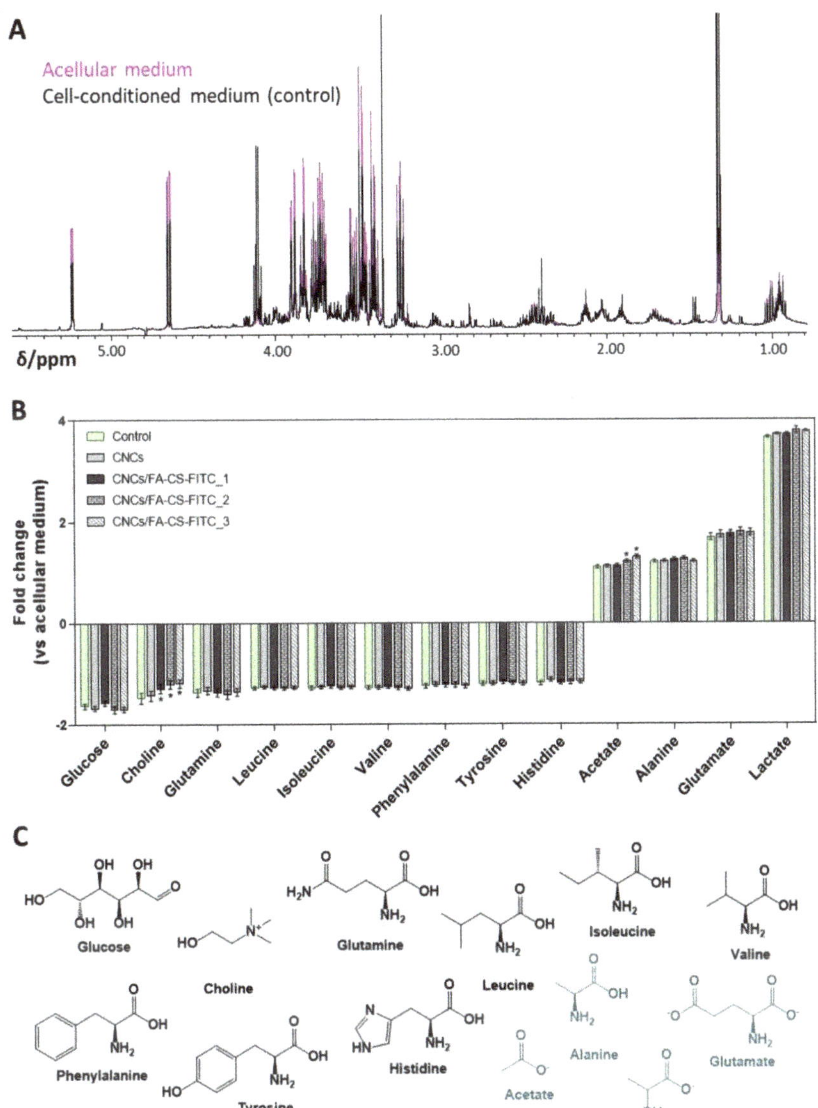

Figure 5. (**A**) ^{1}H NMR spectra of the acellular medium (pink) and supernatant from MDA-MB-231 breast cancer cells grown for 24 h (black), (**B**) variations in metabolites consumed (negative bars) and excreted (positive bars) by the MDA-MB-231 cells, under control conditions and upon treatment with 200 µg mL^{-1} of CNCs, CNCs/FA-CS-FITC_1, CNCs/FA-CS-FITC_2 and CNCs/FA-CS-FITC_3 for 24 h (the symbol * represents the means with a significant difference from the control at $p < 0.05$), and (**C**) chemical structures of the consumed (black) and excreted (green) metabolites.

Henceforth, the CNCs/FA-CS- FITC nanosystems can be exploited as nanocarriers of imaging and chemotherapeutics agents for active targeted therapy. In fact, the incorporation of, for instance, gold nanoparticles (with potential for photothermal cancer therapy [67]) into the CNCs/FA-CS-FITC nanosystems could be an option to engineer a nanocarrier with combined diagnostic and therapeutic capabilities, viz. a theranostic nanosystem [68].

4. Conclusions

In the present work, nanosystems composed of CNCs and a CS derivative were successfully developed and characterized. The physical adsorption of the CS derivative, containing a targeting ligand (FA) and an imaging agent (FITC), on the surface of the CNCs was an eco-friendly methodology to obtain CNCs-based nanosystems. The ensuing nanosystems displayed good stability in two distinct simulated non-physiological and physiological conditions (pH 2.1 and 7.2, respectively), as well as non-cytotoxicity towards MDA-MB-231 cells up to a concentration of 200 µg mL^{-1}. The CNCs nanosystems showed a superior cellular internalization with a 5-fold increase in the fluorescence intensity for the nanosystem with the greater content of FA, viz. concentration-dependent internalization by the FR-positive breast cancer cells. Additionally, the exometabolomics of MDA-MB-231 cells exposed to the CNCs/FA-CS-FITC nanosystems revealed a mild impact on the metabolic activity of the cells, namely a decreased choline uptake and increased acetate release, which suggests an anti-proliferative effect. The overall data evidenced that the elongated CNCs/FA-CS-FITC nanosystems produced by an eco-friendly methodology have high affinity towards folate receptor-positive cancer cells with enhanced cellular internalization, and hence might be employed as nanocarriers with imaging properties for active targeted therapy.

Author Contributions: Conceptualization, C.S.R.F. and C.V.; investigation, R.J.B.P., N.S.L., G.G., G.H.R.d.S., P.O. and C.V.; resources, S.S., H.O., I.F.D. and C.S.R.F.; writing—original draft preparation, C.V. and R.J.B.P.; writing—review and editing, R.J.B.P., N.S.L., G.G., G.H.R.d.S., P.O., S.S., H.O., I.F.D., C.V. and C.S.R.F.; supervision, C.S.R.F.; funding acquisition, C.S.R.F., I.F.D., H.O. and S.S. All authors have read and agreed to the published version of the manuscript.

Funding: This work was developed within the scope of the projects CICECO—Aveiro Institute of Materials (UIDB/50011/2020 & UIDP/50011/2020) and CESAM—Centre for Environmental and Marine Studies (UIDB/50017/2020 & UIDP/50017/2020) financed by national funds through the Portuguese Foundation for Science and Technology (FCT)/MCTES. The research contract of R.J.B.P. was funded by national funds (OE), through FCT in the scope of the framework contract foreseen in the numbers 4, 5, and 6 of the article 23, of the Decree-Law 57/2016, of 29 August, changed by Law 57/2017, of 19 July. FCT is also acknowledged for the doctoral grant to N.S.L. (SFRH/BD/140229/2018) and the research contracts under Scientific Employment Stimulus to H.O. (CEECIND/04050/2017), C.V. (CEECIND/00263/2018) and C.S.R.F. (CEECIND/00464/2017). FAPESP (Fundação de Amparo à Pesquisa do Estado de São Paulo) and the University of Aveiro are acknowledged for the doctoral grants to G.H.R.d.S. (2018/24814-0) and P.O. (BD/REIT/8623/2020), respectively. The NMR spectrometer is part of the National NMR Network (PTNMR), partially supported by Infrastructure Project N° 022161 (co-financed by FEDER through COMPETE 2020, POCI and PORL and FCT through PIDDAC).

Conflicts of Interest: The authors declare no conflict of interest.

References

1. Vilela, C.; Pinto, R.J.B.; Pinto, S.; Marques, P.A.A.P.; Silvestre, A.J.D.; Freire, C.S.R. *Polysaccharide Based Hybrid Materials: Metals and Metal Oxides, Graphene and Carbon Nanotubes*, 1st ed.; Springer: Berlin/Heidelberg, Germany, 2018; ISBN 978-3-030-00346-3.
2. Klemm, D.; Heublein, B.; Fink, H.-P.; Bohn, A. Cellulose: Fascinating biopolymer and sustainable raw material. *Angew. Chem. Int. Ed.* **2005**, *44*, 3358–3393. [CrossRef] [PubMed]
3. Klemm, D.; Cranston, E.D.; Fischer, D.; Gama, M.; Kedzior, S.A.; Kralisch, D.; Kramer, F.; Kondo, T.; Lindström, T.; Nietzsche, S.; et al. Nanocellulose as a natural source for groundbreaking applications in materials science: Today's state. *Mater. Today* **2018**, *21*, 720–748. [CrossRef]
4. Almeida, T.; Silvestre, A.J.D.; Vilela, C.; Freire, C.S.R. Bacterial nanocellulose toward green cosmetics: Recent progresses and challenges. *Int. J. Mol. Sci.* **2021**, *22*, 2836. [CrossRef] [PubMed]
5. Vilela, C.; Silvestre, A.J.D.; Figueiredo, F.M.L.; Freire, C.S.R. Nanocellulose-based materials as components of polymer electrolyte fuel cells. *J. Mater. Chem. A* **2019**, *7*, 20045–20074. [CrossRef]
6. Heise, K.; Kontturi, E.; Allahverdiyeva, Y.; Tammelin, T.; Linder, M.B.; Nonappa; Ikkala, O. Nanocellulose: Recent fundamental advances and emerging biological and biomimicking applications. *Adv. Mater.* **2020**, *33*, 2004349. [CrossRef]
7. Meng, L.Y.; Wang, B.; Ma, M.G.; Zhu, J.F. Cellulose-based nanocarriers as platforms for cancer therapy. *Curr. Pharm. Des.* **2017**, *23*, 5292–5300. [CrossRef]

8.   Ul-islam, S.; Ul-islam, M.; Ahsan, H.; Ahmed, M.B.; Shehzad, A.; Fatima, A.; Sonn, J.K.; Lee, Y.S. Potential applications of bacterial cellulose and its composites for cancer treatment. *Int. J. Biol. Macromol.* **2021**, *168*, 301–309. [CrossRef]

9.   Seabra, A.B.; Bernardes, J.S.; Fávaro, W.J.; Paula, A.J.; Durán, N. Cellulose nanocrystals as carriers in medicine and their toxicities: A review. *Carbohydr. Polym.* **2018**, *181*, 514–527. [CrossRef] [PubMed]

10.   Silva, A.C.Q.; Vilela, C.; Santos, H.A.; Silvestre, A.J.D.; Freire, C.S.R. Recent trends on the development of systems for cancer diagnosis and treatment by microfluidic technology. *Appl. Mater. Today* **2020**, *18*, 100450. [CrossRef]

11.   World Health Organization Fact Sheets: Cancer. Available online: https://www.who.int/news-room/fact-sheets/detail/cancer (accessed on 27 May 2021).

12.   Ferlay, J.; Ervik, M.; Lam, F.; Colombet, M.; Mery, L.; Piñeros, M.; Znaor, A.; Soerjomataram, I.; Bray, F. Global Cancer Observatory: Cancer Today, Lyon, France: International Agency for Research on Cancer. Available online: https://gco.iarc.fr/today (accessed on 27 May 2021).

13.   Trache, D.; Hussin, M.H.; Haafiz, M.K.M.; Thakur, V.K. Recent progress in cellulose nanocrystals: Sources and production. *Nanoscale* **2017**, *9*, 1763–1786. [CrossRef]

14.   Gratton, S.E.A.; Ropp, P.A.; Pohlhaus, P.D.; Luft, J.C.; Madden, V.J.; Napier, M.E.; DeSimone, J.M. The effect of particle design on cellular internalization pathways. *Proc. Natl. Acad. Sci. USA* **2008**, *105*, 11613–11618. [CrossRef]

15.   Blanco, E.; Shen, H.; Ferrari, M. Principles of nanoparticle design for overcoming biological barriers to drug delivery. *Nat. Biotechnol.* **2015**, *33*, 941–951. [CrossRef]

16.   Liu, Y.; Tan, J.; Thomas, A.; Ou-Yang, D.; Muzykantov, V.R. The shape of things to come: Importance of design in nanotechnology for drug delivery. *Ther. Deliv.* **2012**, *3*, 181–194. [CrossRef] [PubMed]

17.   Shang, L.; Nienhaus, K.; Nienhaus, G.U. Engineered nanoparticles interacting with cells: Size matters. *J. Nanobiotechnol.* **2014**, *12*, 5. [CrossRef] [PubMed]

18.   Khine, Y.Y.; Stenzel, M.H. Surface modified cellulose nanomaterials: A source of non-spherical nanoparticles for drug delivery. *Mater. Horiz.* **2020**, *7*, 1727–1758. [CrossRef]

19.   Senapati, S.; Mahanta, A.K.; Kumar, S.; Maiti, P. Controlled drug delivery vehicles for cancer treatment and their performance. *Signal Transduct. Target. Ther.* **2018**, *3*, 7. [CrossRef]

20.   Dong, S.; Roman, M. Fluorescently labeled cellulose Nanocrystals for Bioimaging Applications. *J. Am. Chem. Soc.* **2007**, *129*, 13810–13811. [CrossRef] [PubMed]

21.   Roman, M.; Dong, S.; Hirani, A.; Lee, Y.W. Cellulose nanocrystals for drug delivery. In *Polysaccharide Materials: Performance by Design*; Edgar, K.J., Heinze, T., Buchanan, C.M., Eds.; ACS Symposium Series; American Chemical Society: Washington, DC, USA, 2010; Volume 1017, pp. 81–91.

22.   Dong, S.; Hirani, A.A.; Colacino, K.R.; Lee, Y.W.; Roman, M. Cytotoxicity and cellular uptake of cellulose nanocrystals. *Nano Life* **2012**, *2*, 1241006. [CrossRef]

23.   Dong, S.; Cho, H.J.; Lee, Y.W.; Roman, M. Synthesis and cellular uptake of folic acid-conjugated cellulose nanocrystals for cancer targeting. *Biomacromolecules* **2014**, *15*, 1560–1567. [CrossRef] [PubMed]

24.   Colacino, K.R.; Arena, C.B.; Dong, S.; Roman, M.; Davalos, R.V.; Lee, Y.W. Folate conjugated cellulose nanocrystals potentiate irreversible electroporation-induced cytotoxicity for the selective treatment of cancer cells. *Technol. Cancer Res. Treat.* **2015**, *14*, 757–766. [CrossRef]

25.   Bittleman, K.R.; Dong, S.; Roman, M.; Lee, Y.W. Folic acid-conjugated cellulose nanocrystals show high folate-receptor binding affinity and uptake by KB and breast cancer cells. *ACS Omega* **2018**, *3*, 13952–13959. [CrossRef] [PubMed]

26.   Raja, S.; Hamouda, A.E.I.; De Toledo, M.A.S.; Hu, C.; Bernardo, M.P.; Schalla, C.; Leite, L.S.F.; Buhl, E.M.; Dreschers, S.; Pich, A.; et al. Functionalized cellulose nanocrystals for cellular labeling and bioimaging. *Biomacromolecules* **2021**, *22*, 454–466. [CrossRef] [PubMed]

27.   Tortorella, S.; Buratti, V.V.; Maturi, M.; Sambri, L.; Franchini, M.C.; Locatelli, E. Surface-modified nanocellulose for application in biomedical engineering and nanomedicine: A review. *Int. J. Nanomed.* **2020**, *15*, 9909–9937. [CrossRef]

28.   Babu, A.; Ramesh, R. Multifaceted applications of chitosan in cancer drug delivery and therapy. *Mar. Drugs* **2017**, *15*, 96. [CrossRef] [PubMed]

29.   Liu, Y.; Wang, H.; Yu, G.; Yu, Q.; Li, B.; Mu, X. A novel approach for the preparation of nanocrystalline cellulose by using phosphotungstic acid. *Carbohydr. Polym.* **2014**, *110*, 415–422. [CrossRef] [PubMed]

30.   Lee, D.; Lockey, R.; Mohapatra, S. Folate receptor-mediated cancer cell specific gene delivery using folic acid-conjugated oligochitosans. *J. Nanosci. Nanotechnol.* **2006**, *6*, 2860–2866. [CrossRef] [PubMed]

31.   Huang, Y.; Boamah, P.O.; Gong, J.; Zhang, Q.; Hua, M.; Ye, Y. Gd(III) complex conjugate of low-molecular-weight chitosan as a contrast agent for magnetic resonance/fluorescence dual-modal imaging. *Carbohydr. Polym.* **2016**, *143*, 288–295. [CrossRef] [PubMed]

32.   Tobaldi, E.; Dovgan, I.; Mosser, M.; Becht, J.-M.; Wagner, A. Structural investigation of cyclo-dioxo maleimide cross-linkers for acid and serum stability. *Org. Biomol. Chem.* **2017**, *15*, 9305–9310. [CrossRef]

33.   Caprifico, A.E.; Polycarpou, E.; Foot, P.J.S.; Calabrese, G. Biomedical and pharmacological uses of fluorescein isothiocyanate chitosan-based nanocarriers. *Macromol. Biosci.* **2021**, *21*, 2000312. [CrossRef]

34.   Anitha, A.; Sowmya, S.; Kumar, P.T.S.; Deepthi, S.; Chennazhi, K.P.; Ehrlich, H.; Tsurkan, M.; Jayakumar, R. Chitin and chitosan in selected biomedical applications. *Prog. Polym. Sci.* **2014**, *39*, 1644–1667. [CrossRef]

35. Vilela, C.; Figueiredo, A.R.P.; Silvestre, A.J.D.; Freire, C.S.R. Multilayered materials based on biopolymers as drug delivery systems. *Expert Opin. Drug Deliv.* **2017**, *14*, 189–200. [CrossRef]

36. Foster, E.J.; Moon, R.J.; Agarwal, U.P.; Bortner, M.J.; Bras, J.; Camarero-Espinosa, S.; Chan, K.J.; Clift, M.J.D.; Cranston, E.D.; Eichhorn, S.J.; et al. Current characterization methods for cellulose nanomaterials. *Chem. Soc. Rev.* **2018**, *47*, 2609–2679. [CrossRef] [PubMed]

37. Fonseca, D.F.S.; Carvalho, J.P.F.; Bastos, V.; Oliveira, H.; Moreirinha, C.; Almeida, A.; Silvestre, A.J.D.; Vilela, C.; Freire, C.S.R. Antibacterial multi-layered nanocellulose-based patches loaded with dexpanthenol for wound healing applications. *Nanomaterials* **2020**, *10*, 2469. [CrossRef]

38. Chakraborty, P.; Bairi, P.; Roy, B.; Nandi, A.K. Improved mechanical and electronic properties of co-assembled folic acid gel with aniline and polyaniline. *ACS Appl. Mater. Interfaces* **2014**, *6*, 3615–3622. [CrossRef] [PubMed]

39. Lee, Y.-C.; Lee, T.; Han, H.; Go, W.J.; Yang, J.; Shin, H. Optical properties of fluorescein-labeled organoclay. *Photochem. Photobiol.* **2010**, *86*, 520–527. [CrossRef]

40. Ghosh, S.K.; Ali, M.; Chatterjee, H. Studies on the interaction of fluorescein isothiocyanate and its sugar analogues with cetyltrimethylammonium bromide. *Chem. Phys. Lett.* **2013**, *561–562*, 147–152. [CrossRef]

41. Wang, L.; Roitberg, A.; Meuse, C.; Gaigalas, A.K. Raman and FTIR spectroscopies of fluorescein in solutions. *Spectrochim. Acta Part A Mol. Biomol. Spectrosc.* **2001**, *57*, 1781–1791. [CrossRef]

42. Bellamy, L.J. *The Infrared Spectra of Complex Molecules*, 3rd ed.; Chapman and Hall, Ltd.: London, UK, 1975; ISBN 041-2-138-506.

43. Alupei, L.; Lisa, G.; Butnariu, A.; Desbrières, J.; Cadinoiu, A.N.; Peptu, C.A.; Calin, G.; Popa, M. New folic acid-chitosan derivative based nanoparticles—Potential applications in cancer therapy. *Cellul. Chem. Technol.* **2017**, *51*, 631–648.

44. Elshoky, H.A.; Salaheldin, T.A.; Ali, M.A.; Gaber, M.H. Ascorbic acid prevents cellular uptake and improves biocompatibility of chitosan nanoparticles. *Int. J. Biol. Macromol.* **2018**, *115*, 358–366. [CrossRef]

45. Nawaz, A.; Wong, T.W. Chitosan-carboxymethyl-5-fluorouracil-folate conjugate particles: Microwave modulated uptake by skin and melanoma cells. *J. Investig. Dermatol.* **2018**, *138*, 2412–2422. [CrossRef]

46. Filpponen, I.; Sadeghifar, H.; Argyropoulos, D.S. Photoresponsive cellulose nanocrystals. *Nanomater. Nanotechnol.* **2011**, *1*, 34–43. [CrossRef]

47. Pinto, R.J.B.; Fernandes, S.C.M.; Freire, C.S.R.; Sadocco, P.; Causio, J.; Neto, C.P.; Trindade, T. Antibacterial activity of optically transparent nanocomposite films based on chitosan or its derivatives and silver nanoparticles. *Carbohydr. Res.* **2012**, *348*, 77–83. [CrossRef]

48. Zhao, J.; Wu, J. Preparation and characterization of the fluorescent chitosan nanoparticle probe. *Chin. J. Anal. Chem.* **2006**, *34*, 1555–1559. [CrossRef]

49. Mahmoud, K.A.; Mena, J.A.; Male, K.B.; Hrapovic, S.; Kamen, A.; Luong, J.H.T. Effect of surface charge on the cellular uptake and cytotoxicity of fluorescent labeled cellulose nanocrystals. *ACS Appl. Mater. Interfaces* **2010**, *2*, 2924–2932. [CrossRef]

50. Mohanta, V.; Madras, G.; Patil, S. Albumin-mediated incorporation of water-insoluble therapeutics in layer-by-layer assembled thin films and microcapsules. *J. Mater. Chem. B* **2013**, *1*, 4819–4827. [CrossRef]

51. Marques, M.R.C.; Loebenberg, R.; Almukainzi, M. Simulated biological fluids with possible application in dissolution testing. *Dissolution Technol.* **2011**, *18*, 15–28. [CrossRef]

52. Gorgieva, S.; Vivod, V.; Maver, U.; Gradišnik, L.; Dolenšek, J.; Kokol, V. Internalization of (bis)phosphonate-modified cellulose nanocrystals by human osteoblast cells. *Cellulose* **2017**, *24*, 4235–4252. [CrossRef]

53. Hassan, M.L.; Moorefield, C.M.; Elbatal, H.S.; Newkome, G.R.; Modarelli, D.A.; Romano, N.C. Fluorescent cellulose nanocrystals via supramolecular assembly of terpyridine-modified cellulose nanocrystals and terpyridine-modified perylene. *Mater. Sci. Eng. B* **2012**, *177*, 350–358. [CrossRef]

54. Chavez, K.J.; Garimella, S.V.; Lipkowitz, S. Triple negative breast cancer cell lines: One tool in the search for better treatment of triple negative breast cancer. *Breast Dis.* **2011**, *32*, 35–48. [CrossRef]

55. Meier, R.; Henning, T.D.; Boddington, S.; Tavri, S.; Arora, S.; Piontek, G.; Rudelius, M.; Corot, C.; Daldrup-Link, H.E. Breast Cancers: MR imaging of folate-receptor expression with the folate-specific nanoparticle P1133. *Radiology* **2010**, *255*, 527–535. [CrossRef] [PubMed]

56. Thakur, V.; Kutty, R.V. Recent advances in nanotheranostics for triple negative breast cancer treatment. *J. Exp. Clin. Cancer Res.* **2019**, *38*, 430. [CrossRef]

57. ISO 10993-5:2009(E). *Biological Evaluation of Medical Devices—Part 5: Tests for In Vitro Cytotoxicity*; ISO: Geneva, Switzerland, 2009.

58. Hemraz, U.D.; Campbell, K.A.; Burdick, J.S.; Ckless, K.; Boluk, Y.; Sunasee, R. Cationic Poly(2-aminoethylmethacrylate) and Poly(N-(2-aminoethylmethacrylamide) modified cellulose nanocrystals: Synthesis, characterization, and cytotoxicity. *Biomacromolecules* **2015**, *16*, 319–325. [CrossRef]

59. Meschini, S.; Pellegrini, E.; Maestri, C.A.; Condello, M.; Bettotti, P.; Condello, G.; Scarpa, M. In vitro toxicity assessment of hydrogel patches obtained by cation-induced cross-linking of rod-like cellulose nanocrystals. *J. Biomed. Mater. Res. Part B Appl. Biomater.* **2020**, *108*, 687–697. [CrossRef]

60. Reardon, A.J.F.; Elliott, J.A.W.; McGann, L.E. Fluorescence as an alternative to light-scatter gating strategies to identify frozen-thawed cells with flow cytometry. *Cryobiology* **2014**, *69*, 91–99. [CrossRef]

61. Guerra, Â.R.; Soares, B.I.G.; Freire, C.S.R.; Silvestre, A.J.D.; Duarte, M.F.; Duarte, I.F. Metabolic effects of a Eucalyptus bark lipophilic extract on triple negative breast cancer and nontumor breast epithelial cells. *J. Proteome Res.* **2021**, *20*, 565–575. [CrossRef] [PubMed]

62. Zheng, J. Energy metabolism of cancer: Glycolysis versus oxidative phosphorylation (review). *Oncol. Lett.* **2012**, *4*, 1151–1157. [CrossRef] [PubMed]

63. Gibellini, F.; Smith, T.K. The Kennedy pathway—De novo synthesis of phosphatidylethanolamine and phosphatidylcholine. *IUBMB Life* **2010**, *62*, 414–428. [CrossRef]

64. Glunde, K.; Bhujwalla, Z.M.; Ronen, S.M. Choline metabolism in malignant transformation. *Nat. Rev. Cancer* **2011**, *11*, 835–848. [CrossRef] [PubMed]

65. Liu, X.; Cooper, D.E.; Cluntun, A.A.; Warmoes, M.O.; Zhao, S.; Reid, M.A.; Liu, J.; Lund, P.J.; Lopes, M.; Garcia, B.A.; et al. Acetate production from glucose and coupling to mitochondrial metabolism in mammals. *Cell* **2018**, *175*, 502–513. [CrossRef] [PubMed]

66. Kamphorst, J.J.; Chung, M.K.; Fan, J.; Rabinowitz, J.D. Quantitative analysis of acetyl-CoA production in hypoxic cancer cells reveals substantial contribution from acetate. *Cancer Metab.* **2014**, *2*, 23. [CrossRef] [PubMed]

67. Vines, J.B.; Yoon, J.H.; Ryu, N.E.; Lim, D.J.; Park, H. Gold nanoparticles for photothermal cancer therapy. *Front. Chem.* **2019**, *7*, 167. [CrossRef]

68. Jaymand, M. Chemically modified natural polymer-based theranostic nanomedicines: Are they the golden gate toward a de novo clinical approach against cancer? *ACS Biomater. Sci. Eng.* **2020**, *6*, 134–166. [CrossRef]

MDPI

St. Alban-Anlage 66

4052 Basel

Switzerland

Tel. +41 61 683 77 34

Fax +41 61 302 89 18

www.mdpi.com

Nanomaterials Editorial Office

E-mail: nanomaterials@mdpi.com

www.mdpi.com/journal/nanomaterials